Small Spacecraft Development Project-Based Learning

Jeremy Straub • Ronald Arthur Marsh
David J. Whalen

Small Spacecraft Development Project-Based Learning

Implementation and Assessment of an Academic Program

Jeremy Straub
Department of Computer Science
North Dakota State University
Fargo, ND, USA

Ronald Arthur Marsh
Department of Computer Science
University of North Dakota
Grand Forks, ND, USA

David J. Whalen
Department of Space Studies
University of North Dakota
Grand Forks, ND, USA

ISBN 978-3-319-23644-5   ISBN 978-3-319-23645-2 (eBook)
DOI 10.1007/978-3-319-23645-2

Library of Congress Control Number: 2016953479

© Springer International Publishing Switzerland 2017
This work is subject to copyright. All rights are reserved by the Publisher, whether the whole or part of the material is concerned, specifically the rights of translation, reprinting, reuse of illustrations, recitation, broadcasting, reproduction on microfilms or in any other physical way, and transmission or information storage and retrieval, electronic adaptation, computer software, or by similar or dissimilar methodology now known or hereafter developed.
The use of general descriptive names, registered names, trademarks, service marks, etc. in this publication does not imply, even in the absence of a specific statement, that such names are exempt from the relevant protective laws and regulations and therefore free for general use.
The publisher, the authors and the editors are safe to assume that the advice and information in this book are believed to be true and accurate at the date of publication. Neither the publisher nor the authors or the editors give a warranty, express or implied, with respect to the material contained herein or for any errors or omissions that may have been made.

Printed on acid-free paper

This Springer imprint is published by Springer Nature
The registered company is Springer International Publishing AG
The registered company address is: Gewerbestrasse 11, 6330 Cham, Switzerland

Preface

Small spacecraft, in particular CubeSats, gained significant popularity during the past decade [1]. While space exploration actually began with a small spacecraft (Sputnik [2]), it was only recently that electronics miniaturization and other factors enabled small spacecraft to perform (or even be considered for) missions that were once the domain of much larger spacecraft. A recent news feature in Science [3] contrasted the data collected by a PlanetLabs 10 cm × 10 cm × 30 cm CubeSat and a much larger LandSat spacecraft. Aside from some clouds (the images were taken at different times), the two are virtually indistinguishable. While it would be inaccurate to suggest that small spacecraft can duplicate the capabilities of larger ones in all ways, it is clear that their utility for many applications has been demonstrated.

Despite the value that has been demonstrated to students (see chapter 10) who participate in small spacecraft programs and the research and other capabilities that these spacecraft have provided, very little has been written about the logistics of starting and evaluating a small spacecraft program. Due to this, key questions remain undiscussed. These include what factors should one consider in deciding to start a small spacecraft program, what factors should influence a build vs. buy decision, and how does one evaluate the success of a small spacecraft program. This book seeks to begin to answer some of these questions.

Due to the nature of academic publishing, this book is designed to be read in two ways. The first is as a traditional book: one can start at the beginning and read through areas of interest (skipping and possibly returning to some sections as needed). Many readers, however, may choose to read just a single or small number of chapters. The availability of individual chapters for download (and purchase) via SpringerLink requires that we not assume readers will have read—or even have access to—prior or subsequent chapters. For this reason, a certain amount of background information, required to provide context for the current chapter, is included. Those reading the book straight through will find most of this material presented (in more detail) in Chap. 1 and may wish to skip the background sections in subsequent chapters.

We don't pretend to suggest that the approaches discussed and metrics used in this book are the only ways to start or evaluate a small spacecraft program. In fact, we hope that this work contributes to an ongoing discussion in some areas and starts one in others. We look forward to expanding on this work in the future based on the results of that discussion.

References

1. Swartwout, M. 2014. The first one hundred CubeSats: A statistical look. *Journal of Small Satellites* 2: 213–233.
2. Dickson, P. 2001. *Sputnik: The Shock of the Century.* New York, NY: Walker Publishing Company, Inc.
3. Hand, E. 2015. Startup liftoff. *Science* 348(6231): 172–177. doi:10.1126/science.348.6231.172.

Contents

1	**Introduction**	1
1.1	Overview of Small Spacecraft	1
	1.1.1 Access to Small Spacecraft	2
	1.1.2 The Status Quo	2
	1.1.3 Enter Nanosatellites	3
	1.1.4 Going Mainstream	3
	1.1.5 Academic Proliferation	4
	1.1.6 Start of Industry Proliferation	5
	1.1.7 Comparison of Satellites and Early Computers	5
	1.1.8 The Almost Personal Satellite	6
	1.1.9 Comparison of Capabilities	7
	1.1.10 Towards the Future	7
1.2	Types of Small Spacecraft	8
1.3	Benefits of Small Spacecraft	9
	1.3.1 STEM Education and Small Satellites	9
	1.3.2 Changing Small Satellite Environment	10
	1.3.3 Space Research	10
1.4	Uses of Small Spacecraft	11
	1.4.1 Technologies and Missions	12
	1.4.2 Communications Mission	15
1.5	Conclusion	15
	References	15
2	**Why Start a Small Spacecraft Program**	21
2.1	Overview	21
2.2	Research Benefits	22
2.3	Educational Benefits	24
	2.3.1 Experiential Learning and Problem-Based Learning	24
	2.3.2 Benefits of Interdisciplinary Projects	26
2.4	Societal Benefits	27
	2.4.1 Remote Sensing Benefits, Data Products, and Their Uses	27

	2.4.2	Technologies and Mission	28
	2.4.3	Collaborative Mission for Developing Countries	29
	2.4.4	Qualitative Analysis	29
2.5	Considerations Based on National Space Competency		31
2.6	Conclusion		31
References			31

3 To Build, Buy, or in Between? ... 37

3.1	Why Launch a NanoSat?		37
	3.1.1	Low-Cost Test Platform	38
	3.1.2	Capability to Mature Technical Readiness of Experimental Space Technologies	38
	3.1.3	Ecosystem of Innovation	39
3.2	Overview of Different Approaches to Spacecraft Development		39
3.3	Kit-Based Approach		40
3.4	Bespoke Approach		42
3.5	Framework-Based Approach		42
3.6	Qualitative Evaluation of the Value of the Approaches		42
	3.6.1	Cost Levels	43
	3.6.2	Consideration of Recurring Amortized Vendor Development Costs	48
	3.6.3	Ease of Modification and Extensions of Design	49
	3.6.4	Allowing Focus on Area of Interest	50
	3.6.5	Benefits Related to Export Control (EAR/ITAR)	50
3.7	Mix and Match		51
3.8	Conclusions		51
References			51

4 Starting a Small Spacecraft Program: Types of Programs and Their Benefits and Drawbacks ... 53

4.1	Background		54
	4.1.1	Project-Based Learning and Experiential Education	54
	4.1.2	Small Spacecraft Development	55
4.2	Internal Research Program		55
	4.2.1	Benefits	56
	4.2.2	Drawbacks	56
4.3	External Partner Research Program		56
	4.3.1	Benefits	57
	4.3.2	Drawbacks	58
4.4	Education-Only Program		58
	4.4.1	Benefits	59
	4.4.2	Drawbacks	61
4.5	Hybrid Research Education Program		61
4.6	Academic Institution Decision-Making Process		62
4.7	Summary		62
References			62

5 Forming a Program: Funding and Organizational Issues ... 65
- 5.1 Defining Resource Needs: Human Resources ... 65
  - 5.1.1 Student Involvement ... 66
  - 5.1.2 Student Risk Perception ... 66
- 5.2 Defining Resource Needs: Financial and Other Resources ... 67
- 5.3 Organizational Strategies for Program Formation ... 68
  - 5.3.1 Program Implementation ... 69
  - 5.3.2 Implementation Difficulties ... 72
- 5.4 Conclusion ... 74
- References ... 74

6 Forming a Program: Technical and Logistical Issues ... 77
- 6.1 Considering the Reasons for Forming a Program ... 77
- 6.2 Identifying Science, Technology Development, Educational, and Other Goals ... 78
  - 6.2.1 Defining Objectives ... 78
  - 6.2.2 Maximizing Value ... 79
  - 6.2.3 Value Assessment ... 80
- 6.3 Matching Goals and Funding Sources ... 83
- 6.4 Goal-Based Technique for Requirement and Constraint Decision Making ... 84
  - 6.4.1 Defining Requirements and Constraints ... 84
  - 6.4.2 Creating and Selecting a Mission Concept ... 85
  - 6.4.3 Using Objectives, Requirements, and Constraints for Decision Making ... 86
- 6.5 Picking a Design Framework and the Level of Program Rigidity ... 86
  - 6.5.1 Comparison of Design Frameworks ... 86
  - 6.5.2 Mission Analysis and Design ... 88
  - 6.5.3 Defining the Mission Architecture ... 90
  - 6.5.4 Driver Identification ... 91
  - 6.5.5 Requirements, Analysis, and Selection ... 91
- 6.6 Planning for Program Longevity: Technical and Logistical Considerations ... 92
  - 6.6.1 Conceptualization ... 92
  - 6.6.2 Design ... 93
  - 6.6.3 Development ... 93
  - 6.6.4 Launch and Operations ... 93
  - 6.6.5 Closeout ... 94
- 6.7 Processes for Mission Management ... 94
  - 6.7.1 Project/Mission Management ... 94
  - 6.7.2 Systems and Processes ... 95
  - 6.7.3 Assurance ... 96
- 6.8 Conclusion ... 96
- References ... 97

7	**Student Involvement and Risk**	101
7.1	Introduction	101
7.2	Background	102
	7.2.1 Project-Based Learning	102
	7.2.2 Value of Student Involvement to Faculty Research	103
	7.2.3 Risk Perception	103
7.3	The Student Qualitative Undertaking Involvement Risk Model	104
	7.3.1 Technical, Schedule and Other Standard Risks	104
	7.3.2 Technical Risk	105
	7.3.3 Schedule Risk	106
	7.3.4 Cost Risk	108
	7.3.5 Risks Posed by Student Worker Involvement	109
7.4	Extending the Model with Root Cause Analysis Techniques	110
	7.4.1 Root Cause Analysis	112
	7.4.2 Inexperience Symptoms Occur	112
	7.4.3 Unscheduled Turnover Occurs	114
	7.4.4 Scheduled Turnover Occurs	115
	7.4.5 Miss-commitment	116
7.5	Differences Between and Choosing Between Using SQUIRM and SQUIRM-E	117
	7.5.1 Discussion of the Differences Between SQUIRM and SQUIRM-E	117
	7.5.2 Comparative Simplicity	118
	7.5.3 Types of Inexperienced Workers	118
	7.5.4 Project Size	118
	7.5.5 Familiarity with Particulars of Student Work Environment	118
	7.5.6 Choosing a Model	119
7.6	Application	119
7.7	Quantifying the Model	120
	7.7.1 Risk Assessment	120
	7.7.2 Mitigation/Response Assessment	121
	7.7.3 Combining for Result	121
	7.7.4 Data for Model Parameters	121
7.8	Value Model for Inexperienced Workers	122
	7.8.1 Cost of Inexperienced/Student Workers	122
	7.8.2 Training Benefits	122
	7.8.3 Discontinuous Innovation Benefits	123
7.9	Discussion of the Differences Between Student Volunteers, Paid Student Workers, Interns, and Junior Employees	123
	7.9.1 Student Volunteers	123
	7.9.2 Paid Student Workers	124
	7.9.3 Interns	124
	7.9.4 Junior Employees	124
7.10	Conclusions and Future Work	124
	References	125

8 Setting Educational Goals and Formative Assessment ... 129
- 8.1 Educational Benefits: Overview ... 129
- 8.2 Student Involvement, Faculty Research, and Risk ... 133
- 8.3 Determining Whether the Project Will Be an Educational Program or a Program with Educational Benefits ... 133
  - 8.3.1 Educational Programs ... 133
  - 8.3.2 Programs That Provide an Educational Benefit ... 134
  - 8.3.3 Dual-Objective Programs ... 134
- 8.4 Setting Educational Goals: Technical Discipline Skills ... 135
- 8.5 Setting Educational Goals: Nontraditional Disciplines ... 136
- 8.6 Setting Educational Goals: 'Soft' and Other Skills ... 137
- 8.7 Formative Assessment: Assessing Students' Reasons for Participating and Using This Information ... 138
- 8.8 Formative Assessment: Assessing Whether Goals Are Being Met ... 143
  - 8.8.1 Data Collected ... 145
- 8.9 Incorporating the Results of Formative Assessment ... 146
- 8.10 Conclusion ... 147
- References ... 147

9 Summative Assessment ... 151
- 9.1 Overview ... 151
- 9.2 Background ... 152
  - 9.2.1 Project-Based Learning and Experiential Education ... 152
  - 9.2.2 Small Spacecraft Development ... 152
  - 9.2.3 Assessment ... 153
- 9.3 Small Spacecraft Programs and Their Goals ... 154
- 9.4 Determining Program Value ... 155
  - 9.4.1 Context of the Summative Assessment Process ... 156
  - 9.4.2 Undergraduate Research Student Self-Assessment (URSSA) Mechanism ... 156
  - 9.4.3 Experiment Implementation ... 157
- 9.5 Tracking and Reporting Program Value over Time ... 157
- 9.6 Reporting This Value ... 158
  - 9.6.1 Local Reporting ... 158
  - 9.6.2 Comparison to Other Programs ... 159
- 9.7 Explaining Why the Program Is Important ... 167
- 9.8 Conclusion ... 168
- References ... 168

10 Results of Prior Assessment Work and Its Utility ... 175
- 10.1 Discussion of the Use of Formative Evaluation ... 175
- 10.2 Formative Evaluation Tools: Interest/Reason for Participation Survey ... 176
  - 10.2.1 Experimental Design ... 176
  - 10.2.2 Data Collected ... 178

	10.2.3	Analysis of Data	182
	10.2.4	Summary	183
10.3	Formative Evaluation Tools: Gain Quantification Survey		183
	10.3.1	Benefits of Interdisciplinary Projects	183
	10.3.2	Learning Objectives	184
	10.3.3	Data and Analysis	185
	10.3.4	Overall Results	185
	10.3.5	Comparison of Results Between Undergraduate and Graduate Students	187
	10.3.6	Comparison of Results Between Team Leads and Participants	189
	10.3.7	Comparison of Results by Level of Weekly Participation	191
	10.3.8	Comparison of Results by Amount of Time Participating	193
	10.3.9	Comparison of Results by GPA	195
	10.3.10	Comparison of Results by Undergraduate Class Level	197
	10.3.11	Summary	198
10.4	Formative Evaluation Results and Program Enhancements in Prior Work		198
10.5	Discussion of the Use of Summative Evaluation		199
10.6	Summative Evaluation Tools: Gain Quantification Survey		199
	10.6.1	Results and Discussion	200
	10.6.2	Summary	203
10.7	Summative Evaluation Tools: URSSA		204
	10.7.1	Undergraduate Research Student Self-Assessment Mechanism	205
	10.7.2	Experiment Implementation	205
	10.7.3	Summary	214
10.8	Summative Evaluation Tools: Meeting Expectations Survey		214
	10.8.1	Data Collected	216
	10.8.2	Analysis of Data	219
	10.8.3	Summary	221
10.9	Results of Prior Summative Evaluation		221
10.10	Conclusion		222
References			222
Index			223

Chapter 1
Introduction

This book explores the formation of small spacecraft development programs. It discusses what factors may lead to program formation, what benefits can be expected from initiating a small spacecraft program, and how to assess those benefits (with a particular focus on educational benefits). Before delving into these topics—and presuming that those considering starting such a program may have limited knowledge about small spacecraft and their development—background information is required. This chapter provides an introduction to and overview of themes that will continue throughout the book. It begins with an overview of small spacecraft. Then, a brief summary of the types of small spacecraft is presented. Next, the benefits of small spacecraft are briefly discussed (a more detailed discussion of this topic is presented in Chap. 2). Finally, an overview of uses of small spacecraft is presented (this topic is covered in greater detail in Chaps. 3 and 4).

1.1 Overview of Small Spacecraft

Access to space is still largely limited to national governments, large corporations, and those with the support of the foregoing. Spacecraft design, development, and launch costs are a barrier to entry for most that might be otherwise interested. This is a problem for everyone from would-be commercial craft operators to educators to hobbyists. There is a large community that wants to interact with space—unfortunately, most of them don't get the chance. While small satellite form factors, such

---

This chapter is based on, revises, and extends the papers "CubeSats: A Low-Cost, Very High-Return Space Technology" [1], "Increasing National Space Engineering Productivity and Educational Opportunities via Intrepreneurship, Entrepreneurship and Innovation" [2], "Evaluation of the Educational Impact of Participation Time in a Small Spacecraft Development Program" [3], and "Application of Collaborative Autonomous Control and the OPEN Prototype for Educational NanoSats Framework to Enable Orbital Capabilities for Developing Nations" [4].

as the CubeSat, have made access more attainable, it is still not within the reach of most companies and universities, and virtually all hobbyists.

A larger community, however, is required to drive costs down and make space more universally accessible. The involvement of lower-cost launch providers, small and mid-sized businesses, students, educators, hobbyists, and others is precisely what is needed to start a downward spiral of costs and upward spiral of access. A comparison of the satellite and early computer industry shows how an expanded base of interested and involved individuals can dramatically increase access to space.

1.1.1 Access to Small Spacecraft

In the early days of computers, one had to work for a government agency, a top-tier research institution, or one of the largest corporations in order to have access to a computer. Computers were run by highly trained operators in specialized facilities. There were limited people who understood them—and a specialized few that could fix them when they broke. Over the course of the next 70 years, global computational capabilities increased exponentially. Computing technology moved from being housed in specialized facilities to being housed in the pockets, houses, and offices of a significant proportion of the population. In fact, the decentralized computing capabilities have expanded to the point where researchers (the people who might have had access to the large secured computers) are attempting to make use of the excess capabilities of personal and office computers to perform supercomputer grade research (e.g., [5]).

Paralleling early computing, the earliest spacecraft were created by government (or by industry for government). These spacecraft were very large and heavy; their performance was constrained by the size and weight of their components.

1.1.2 The Status Quo

In the past 55 years, not much has changed. Satellites have gotten somewhat smaller (or more powerful, at a given size); however, they are still largely the province of government or large businesses. Between 1999 and 2009, the average cost of a satellite (excluding those under 40 kg and government-classified satellites) was $97 million; launch costs averaged $51 million [6]. Satellite costs are projected to increase to $99 million for the 2009–2019 period; launch costs were not expected to vary significantly [6]. A $150 million price tag places access to space outside the consideration of small and medium-sized businesses, most researchers, and even many governments.

The high price also creates risk aversion. This leads to the incorporation of redundancy in satellite systems (which contributes to cost) and also drives the use of highly tested technologies, which may be two or more generations behind the capabilities of the leading edge. One can look at this as possibly forming a vicious cycle where space capabilities fall further and further behind terrestrial capabilities—while costing orders of magnitude more.

1.1.3 Enter Nanosatellites

Another problem posed by the cost and risk aversion of the space engineering industry is entry. How could students gain the skills to work on these large and expensive spacecraft, when it was highly desirable that those working on the projects have significant space engineering experience? In response to this problem, Bob Twigs and Jordi Puig-Suari proposed the creation of small cube-shaped satellites, in 1999 [7].

As component technologies were further miniaturized, however, it became apparent that these small satellites could do far more than provide an educational experience for students. Their light weight (and associated low launch costs) makes them a desirable platform for numerous mission concepts, including technology demonstration [8, 9], scientific investigation [10–13], need-responsive remote sensing [14], and possibly even deep-space exploration [15, 16].

The cost of CubeSats is decreasing. According to Twiggs and Malphrus, initial projects required budgets in the neighborhood of $250,000 [7]. However, growing interest and the entry of multiple firms has dramatically reduced these costs, with up-and-coming launch provider Interorbital Systems offering to provide a skeleton kit and a launch to 310 km for $19,125 [17]. They also offer a slightly smaller (75% of the volume) tube-shaped kit and 310-km orbital launch for $8000 [18]. They stated that they would begin launches in 2012 [18]; however, this has not yet occurred. Other firms, such as Tyvak, offer a more complete kit for approximately $40,000 [19]. Several approaches were compared by Berk et al. [20] which have cost levels of $5000 or below (in terms of parts only). While both the Interorbital and Tyvak offerings would have additional costs and require some labor to build and integrate the craft's payload and other required components, the existence of these skeleton systems reduces risk and cost. This is demonstrated by the fact that a CubeSat project has even been undertaken at a US-based high school [21].

1.1.4 Going Mainstream

The changing budgetary and operational environment has triggered an interest in small satellites from a number of additional sectors. Military and intelligence services are looking at smaller-sized satellites to create a highly responsive, launch-on-demand constellations. DARPA's SeeMe program, for example, seeks to use 45 kg or lighter spacecraft to provide the capability to image a target area within 90 min of request [22]. The Army's SMDC-ONE [23] and the Air Force's SMC/XR SENSE [24] satellite concepts are designed to demonstrate small spacecraft's efficacy in meeting operational objectives. Spacecraft of the 3-U CubeSat size (approximately 30 cm × 10 cm × 10 cm, 4 kg) and 6-U size (approximately 60 cm × 20 cm × 10 cm, 8 kg) are seen, by some, as possible candidates for various military and intelligence missions.

On the science side, several efforts highlight the utility of small spacecraft for in situ atmospheric research. The European QB50 program aims to launch 50 2-U CubeSats (approximately 20 cm × 10 cm × 10 cm, 2.66 kg) into orbit to study the lower thermosphere [25]. It is a mission uniquely suited for CubeSats, due to the high number of craft required to carry out the research and short operational lifetime [25]. Risk, organizers state, is mitigated through the number of craft—instead of the traditional onboard fault tolerance and redundancy [25].

The U.S. National Science Foundation has also embraced the use of CubeSats for scientific purposes. The CubeSat-based Space Missions for Geospace and Atmospheric Research program is poised to fund and launch several small spacecraft to perform various proposer-advocated scientific missions [11]. The previous DICE program, a two CubeSat mission to measure ionospheric plasma density and electric fields, represents the first time that a satellite's construction and launch have ever been fully funded by the NSF [13, 26].

While it would be inaccurate to state that CubeSats have already joined their larger cousins in the performance of critical national security, communications, and other similar applications, the aforementioned may form the beginning of a trend that could conceivably lead to this. Additional component miniaturization, space qualification, and numerous successful missions could result in widespread small satellite acceptance.

1.1.5 Academic Proliferation

Driven by lower costs and, undoubtedly, a desire to peak student interest [27], numerous small spacecraft development efforts are underway. It is estimated that 50 CubeSat-class spacecraft have been launched to date and up to 150 could be launched over the next several years [25]. CubeSat programs span at least 23 US states and numerous countries worldwide [28]. This proliferation has given rise to over ten vendors to specifically serve this nascent community [29, 30]. Numerous existing businesses provide components, assemblies, or services to the community. There are at least three distinct commercially available platforms (Pumpkin/PC104, Tyvak, and Interorbital) and two deployer form factor standards (PPOD and Interorbital's IOS Deployment Unit) [17, 31, 32].

CubeSat programs have evolved from being student-engineering exercises and craft flight heritage-building missions [7] to now include performing leading edge scientific and engineering work [33]. Recent efforts have even included looking at the possible use of CubeSat-class spacecraft for interplanetary missions [34].

1.1.6 Start of Industry Proliferation

The most commercialized area of space, satellite communications, was largely developed by industry—not government. In prior work [35] it was demonstrated that industry will only seek to embrace and fund the development of new technology when it sees a clear path to competitive advantage [35]. Given this, it is quite telling to see both large and small companies investing in small satellite technologies.

One of the oldest and best recognized commercial vendors of CubeSat hardware is Pumpkin. Pumpkin Incorporated was formed in 1995 with an initial focus of providing services to the car racing industry [36]. In the past 17 years, the company has been a primary supplier for at least ten CubeSat missions and supplied components for numerous more [37]. Tyvak Nano-Satellite Systems is building several commercial models, building on the Cal Poly spacecraft hardware legacy [38].

There are at least several additional examples of small and mid-sized businesses investing in CubeSat development. In Europe, Clyde Space, Innovative Solutions in Space, and GOMSpace all provide components that are not regulated by US technology transfer regulations [29]. Stras Space, located in Canada, is also less constrained by these regulations [29]. Other companies, such as IntelliTech Microsystems, Princeton Satellite Systems, Space Micro, and Mars Space provide specialized components, tools, or services that facilitate CubeSat development efforts [29].

Also telling is the burgeoning interest in the CubeSat technology from larger entities. One example of this is Boeing. The company, a large defense supplier, in addition to its well-known commercial aircraft business, will be delivering up to 50 3-U CubeSat base models to the National Reconnaissance Office, based on a 2010 order [39]. The company has developed a platform entitled "Colony" [39]. The Colony 1 base units have housed a weather satellite and other classified missions [39]. The Colony 2 model, which enhances pointing accuracy and other craft subsystems, was the subject of the 2010 NRO order [39].

1.1.7 Comparison of Satellites and Early Computers

Like the first satellites, which were developed in response to cold war pressures, the development of the first computers were also a product of national security needs. Campbell-Kelly and Aspray proffer that this computer, developed as part of a project started at the Moore School at the University of Pennsylvania, was built to enable activities at the U.S. Army's Aberdeen Proving Ground [40]. The first two computers to be completed were built in England due to budgetary pressures forcing them to keep their designs simpler than concurrent US-based projects [40].

In the late 1040s and early 1950s, commercialization of computing began with 30 US firms and 10 British firms entering the computer industry [40]. Some, like IBM, came from the predecessor analog counting machine industry [40]. The computers that these firms sought to develop were large and expensive—an analog to the first commercial satellites. However, over the next 20-some years, the components required for making computers decreased in size and cost due to technical innovation and increased demand. By the mid-1970s, scientists, workers, and enthusiasts were able connect to a remote computer using a terminal and purchase processing time for $10–$20 per hour [40].

In January, 1975, the computer industry changed dramatically with the announcement of the Altair 8800—the first computer priced low enough that it was within the purchasing power of individuals [40]. In 1977, computing was made more accessible (and according to Campbell-Kelly and Aspray "the personal computer arrived in the public consciousness") through the launch of the Apple II and Commodore PET; Tandy also launched the TRS-80 later in that year [40].

From here, business and personal computing has grown dramatically. In the United States, now, virtually every office worker has a personal computer on his/her desk. Laptops allow us to perform productivity, calculation, and scientific tasks away from our desks. Tablets, smart phones, and wearable computers allow us to check e-mail, perform calculations, and do numerous other tasks virtually anywhere, any time. Computing is now pervasive.

It is, of course, impossible to know what would have happened had the computing industry continued the paradigm of providing terminal access to large machines. On one hand, the Internet- and network-based services that are currently gaining industry traction are, in some ways, a return to this model. However, an argument can also be made that a lack of access to the physical hardware would have removed the enthusiasm of many of the early developers. It certainly would have made software development more problematic, as developers would have had to contend with the risk aversion of providers who needed to maintain uptime levels for their commercial and government customers.

1.1.8 The Almost Personal Satellite

An analogy can be drawn between the first small satellites and the Altair 8800. They demonstrated technical capabilities and little more. However, more recent small satellites have performed real science and engineering demonstrations that focus on advancing the state of the art.

So how close are CubeSats to being the analog of the personal computer? There is no doubt that that they are far beyond the means of most individuals. Even with a $50,000 price point [17, 19] for a fully functional 1-U CubeSat and launch services (which neglects the costs of actually building the payload components), there is little chance of the United States becoming a nation of satellite owners. The $5000 price point targeted by the Open Prototype for Educational Nanosats design [20] makes this more possible, but still unlikely due to launch and other costs. Of course, the average individual doesn't need to own a satellite (and low-Earth orbit doesn't need the additional traffic that would be generated).

Small satellites provide several pathways towards greater public participation in space, however. First, they facilitate the training of students who will become the next generation of spacecraft engineering professionals—for both small and large

1.1 Overview of Small Spacecraft

spacecraft [41]. The lower-cost, lower-risk platform allows students to participate in a way that simply wouldn't be possible on more risk-adverse missions. The diminished risk aversion also facilitates missions that try out new technologies, concepts, and even new paradigms.

Second, the lower cost facilitates the entry of small and mid-sized businesses into lead project roles and as satellite owners. These businesses lack the capitalization to build a multimillion dollar satellite; however, a satellite with a $100,000–$150,000 price may be affordable.

Third, small satellites facilitate the use of orbital remote sensing for smaller-scale research projects. This allows a project's investigator to choose the best dates and times to collect the most meaningful data (e.g., matching the dates that in situ validation may be scheduled for—or when an important phenomena is expected to occur) as opposed to being limited to commercially produced data. This control also, prospectively, allows the scientist to task the satellite to focus on a not-previously scheduled phenomena of interest, should one be discovered during the research.

Finally, by making small satellites more accessible to the aforementioned—and others—a user/owner base is being formed. Based on the history of most electronic devices, this proliferation should result in even lower prices, making satellite ownership and access available to even more institutions. The formation of a negative cost spiral and positive ownership/access spiral is possible and highly desirable, from a number of prospectives.

1.1.9 Comparison of Capabilities

A key enabler of small satellite proliferation is miniaturization. A recent mission which, in some ways, duplicated a much earlier mission provides an antidotal example of this. In 2011, Montana State University launched the Explorer One Prime Satellite [42]. This 1-U CubeSat was designed to "replicate the scientific mission" of the Explorer One spacecraft, launched in 1958, which detected the Van Allen Radiation Belt [42]. The 1958 model weighed 14 kg and had a dedicated launch, the 2011 satellite weighed less than 1.33 kg and shared a ride with other craft [42, 43]. This order-of-magnitude mass reduction demonstrates why some missions that were previously the domain of larger spacecraft can now be performed by smaller ones. Another example of this is the Stanford HiMARC system [44].

1.1.10 Towards the Future

Small Satellites are poised to promote two complimentary paradigm changes. First, from a technical perspective, they facilitate collaborative missions involving multiple craft. These missions can be less risk averse and launched in response to specific immediate needs. This concept is discussed more in Sect. 1.4.1.3.

From an economic and space industry perspective, small satellites open the proverbial door. Space projects have been shown to be an excellent way to interest students in STEM disciplines [3, 45, 46]. By placing spacecraft technology projects within the reach of universities, state space consortia, and even individual K-12 school districts to sponsor, the United States can truly become a nation of spacefarers (by robotic proxy, at least). The skills that are gained in working with small spacecraft are highly transferable to robotics and a plethora of related disciplines.

By generating interest in—and more importantly—involvement with space, these projects can help shape the national interest. They can, perhaps, form the consensus and national drive that is required to sustain an interplanetary manned space program.

Of course, like any historically based analysis and projection, there are problems with the computer industry analogy. Computers found 'killer apps' that may not exist for spacecraft. There are logistical, regulatory and other hurdles that will prevent the proliferation of satellites from reaching anywhere near the level of computer proliferation in the foreseeable future. Orbital space and communications availability constrain the number of satellites that can effectively operate in orbit at once. A desire to minimize the level of 'space junk' (see, for example, [47]) may result in regulatory actions that are not favorable to the proliferation—particularly if, at some point, a small satellite is involved in a collision with (or otherwise impairs) a larger and more expensive one.

1.2 Types of Small Spacecraft

With the benefits of small spacecraft discussed, it is now time to consider what exactly is a small spacecraft. Fortunately, Wertz presents a taxonomy for the classification of small satellites. He defines smallsats "broadly" as having a mass of less than approximately 500 kg [48]. It should be noted that the precise definition of small satellites remains elusive and, in fact, was a key point of discussion in the formation of the AIAA Small Satellite Technical Committee (which remains unresolved, as of this writing). Using standard prefixes, several subclassifications are also presented including PicoSat (0.1–1 kg), NanoSat (1–10 kg), MicroSat (10–100 kg), and MiniSat (100–500 kg) [48].

Small satellites can also be logically classified by their mission. Some are developed for primarily educational purposes (e.g., [49, 50]); others are developed to support bona fide research [10, 51]. Small spacecraft have also been utilized to support Earth imaging [52], military [23, 24, 53], engineering testing and demonstration [8, 44, 54], and communications purposes [55, 56].

Swartwout [57], however, proposes an alternate factor for classification: whether a mission is "university class" or not. He proposes that university-class missions are able to push boundaries, due to being able to define their success in educational (as opposed to scientific, engineering, or other) goals. To qualify as a university-class mission, the spacecraft must be a free-flier (a deployed spacecraft operating independently), have training as an important (perhaps more important than other mission goals) mission goal, and be developed and constructed by students. Swartwout [58–61] has tracked the extensive growth of the small spacecraft, documenting, in the process, significant growth in both CubeSats and university-class spacecraft.

1.3 Benefits of Small Spacecraft

In 2000, Bob Twiggs dramatically changed the perception of how small a satellite could be through the Orbiting Picosatellite Automatic Launcher (OPAL), which deployed six satellites which had similar dimensions of a hockey puck [62]. The first CubeSat was deployed shortly after this, in 2003, via the Poly-PicoSatellite Orbital Deployer (P-POD) [62]. Since then, more capable CubeSat-class spacecraft and even smaller spacecraft (such as Twiggs' 5 cm×5 cm×5 cm PocketQub and the pocket-size spacecraft proposed by Johnson and others) have been developed [62].

Interplay exists between educational and research uses for small spacecraft. Thakker and Swenson [63] have suggested that the focus, historically, has been primarily on education. Swartwout [64], on the other hand, has suggested that university programs have moved away from being predominately student engineering exercises or "beepsats" (which lack a "compelling" purpose) and can instead serve as "disruptive" platforms for research activities [57] by taking advantage of the greater risk tolerance, student enthusiasm, and innovative ideas present in a university environment. The developers of all of these spacecraft seek benefit in multiple areas which will now be discussed.

1.3.1 STEM Education and Small Satellites

The STEM education benefits from small satellite programs have been well demonstrated. Most CubeSat missions [64] have been university projects involving students. In one example, Thakker and Swenson [65] discuss the organization of CubeSat programs for achieving student benefits. The University of Illinois at Urbana-Champaign's ENGR 491 interdisciplinary design class is presented as an example for emulation. This course, based on a project-based learning (PBL) methodology, involves three faculty advisors, two graduate student project managers (who are teaching assistants for the course), and six teams comprising three to five students enrolled in the course.

Project-based learning is critical to engineering (and related discipline) education; Zhou [66] proffers that it can help drive students to learn to form creative solutions to problems like those that they will later encounter in a workforce environment. Crawley et al. [67] developed a four-step process for PBL which begins with conception and proceeds through design, implementation, and operations. Smith et al. [68] demonstrated the utility of PBL and Crawley's Conceive-Design-Implement-Operate (CDIO) process in the context of CubeSat development. The existence of several long-running small satellite programs (e.g., the University of Hawaii's program [69]) supports the assertion of the value of using PBL for teaching spacecraft development and the value of using spacecraft development to teach skills in component engineering disciplines. The use of students on research projects and projects to teach students is not without risk, however, as many potential sources of potential issues exist [70].

Prior work has demonstrated the efficacy of small spacecraft programs for providing educational benefits. This has been assessed from several different perspectives. First, overall benefit was shown [45]. This was demonstrated by movement of as much as 17% of the total scale in the category of spacecraft design skills, based on self-assessment, which was attributed to participation. Prior work also considered the impact of participation as a function of time [3], specifically for undergraduate students [46] and for those pursuing degrees in the field of computer science [71]. Additional discussion of small satellites and STEM education (as well as some of the aforementioned and other prior work) is presented in Chaps. 7, 8, and 9.

1.3.2 Changing Small Satellite Environment

Woellert [72] proffers that "only a few years ago, one would risk their credibility if they suggested the CubeSat was a viable platform for interplanetary missions"; this, however, is now being seriously researched with several respected conferences focusing on this topic (including CalPoly's CubeSat Workshop and the AIAA/USU Conference on Small Satellites). A few years before this, the notion of performing a bona fide science mission with a CubeSat would be questioned; now their utility for this is generally accepted.

Like many maturing fields, the costs and the barriers to entry to CubeSat construction have declined somewhat. Vendor kits make entry available to anyone with suitable funding [1] and the capability to perform payload integration, and testing. This stands in contrast to earlier missions for which design from the ground up was a necessity. On the spacecraft side, the initial barrier to entry has fallen considerably: from a cost of $250,000 [1] to develop a CubeSat from scratch (and a requirement to have access to specialists in all required areas) to $40,000 plus payload hardware, integration and testing costs [1]. On the launch side, costs are also declining. Vendors have projected costs as low as $10,000 [1] for a 1-U, 1 kg CubeSat launch into low-Earth orbit, as opposed to the $50,000 or more that some missions have paid previously. For educational institutions and nonprofits, a launch can potentially be obtained at no cost to the developer from the NASA ELaNa program [73]. The growing acceptance and proliferation of CubeSats is expanding the number launch vehicles that they can obtain launch services on and the number of orbits that can be reached.

1.3.3 Space Research

Small spacecraft have been demonstrated to provide significant benefit from performing bona fide research activities [7]. Examples of these include engineering development and testing activities such as the use of plastic printed structures, deployable solar panels, advanced (including 3d-printed) propulsion technologies, and structural joints (such as Stanford's HATTS [8]). They have also been used to collect data in support of scientific exploration. Examples of instruments carried as payloads within CubeSats include an oxygen airglow photometer, neutral hydrogen photometer, Langmuir plasma probe, electric field boom, VLF receiver, SSD spectrometer, transient photometer, Langmuir plasma probe, tether, and Nitol tether [63]. Chirayath [44] has discussed their utility for high resolution imaging as part of a multi-CubeSat constellation. In each case, developers have benefited from the satellite's small size, low mass, and significant incorporated capabilities. More uses of small satellites are discussed in Sect. 1.4.

1.4 Uses of Small Spacecraft

Small spacecraft come in many varieties. In fact, the exact definition of the term small spacecraft is elusive. Prefixes have been defined [74] to classify types of spacecraft; however, there is no universally accepted line defining where 'small' ends and larger sizes begin.[1] Swartwout [57, 75], however, proffers that size isn't the defining attribute. Instead, he suggests that the so-called university-class spacecraft should be defined by their educational missions, risk tolerance, and the ability to serve as a testing bed for out-of-the-box concepts. The CubeSat is one form factor that is commonly used for university-class spacecraft. Developed initially by Bob Twiggs and Jordi Puig-Suari as a tool to facilitate aerospace engineering education [62], CubeSats are now widely used by education [59, 76] as well as being developed for science [10, 44, 51, 77], government [78], military [24, 53], and commercial [9, 38] purposes. Their development is being aided [1] by the availability of free-to-qualified-developer launch services from the U.S. Air Force [79], NASA [73], and the ESA [80]. Reduced cost commercial launches are also on the horizon [81, 82]. Low-cost development approaches, such as OPEN [20], are also enabling adoption via reducing the cost of spacecraft development. In 2013, 30 academic and 50 nonacademic CubeSats were manifested and over 100 institutions had participated in the development of a CubeSat-class spacecraft [59]. Over 80 were actually launched [61]. In 2014, 80 CubeSats were launched and nearly 90 spacecraft lighter than 10 kg were launched. Overall, more than 110 secondary spacecraft were launched in 2014.

A wide variety of data products can be generated via remote sensing (including thermal infrared imaging, multispectral imaging, microwave and LIDAR imaging, and gravitational data [83]). Historically, visible light imagery (and near-visible light imagery that can be produced via changing the filtering applied to standard sensing equipment) has been a prevalent remote sensing technique.

Visible light remote sensing data is defined by its coverage and spatial and temporal resolution as well as other qualitative aspects [83]. Consideration of all of these factors is critical in assessing the suitability of a sensing system. Coverage of

---

[1] This was a significant topic of discussion at the formation meeting of the AIAA Small Satellite Technical Committee that remains unresolved.

desired areas is clearly important as is spatial resolution (a measure of the size represented by each pixel on the imagery). The utility of spatial resolution levels ranging from submeter to over a kilometer has been demonstrated [84]. Temporal resolution is a measure of how frequently data for a given area can be reobtained (e.g., how current the data product is).

For agriculture, visible light sensing data can be useful for assessing where to deploy and the deployment of fungicides and pesticides, assessing crop damage due to weather and assessing drainage patterns and designing drainage solutions [85]. The utility of both aerial and satellite imagery has been demonstrated for this purpose [83, 85, 86]. The resolution required varies by application; however, the utility of 20-m data has been demonstrated by the International Space Station Agricultural Camera [87]; the use of 10-m [85] and much higher resolution [86] data has also been demonstrated. Temporal requirements vary, based on the desired phenomena under study; in addition, data may be needed on demand for use in storm damage assessment and other cases.

Remote sensing of urban areas, used for municipal, county, and state-level planning and other purposes, has been performed with data ranging from 100-m to 10-m or higher resolution [88]. This data can be used to determine material composition, land cover, and land use for planning and other purposes. The level of temporal coverage required varies significantly, with once-a-year resolution being acceptable for some applications and more frequent imagery (including imagery at given times) being required for others. Miller and Small [89] proffer that remote sensed imagery is particularly important for developing regions, as it can serve to replace the growth and environmental condition data collected in situ (or by other means) in more developed nations. They also note the utility of remotely sensed data being unobstructive and consistent.

Remote sensed data has also been shown to be useful for the response to hazards such as earthquakes, volcanos, floods, landslides, and costal inundations. In this context, the data can be used to prioritize response efforts, direct responders as well as to, in the longer term, perform risk assessments, and take actions and establish policies to prevent future issues [90].

A multitude of other uses for remote-sensed data exist, many of which would be relevant to developing nations. These include its use in aquaculture (sea farming), such as was demonstrated in India [91], and water policy [92].

1.4.1 Technologies and Missions

For purposes of illustration, a multi-craft remote sensing mission is now considered. Several technologies are required to support such a mission. In addition to a capable spacecraft bus and standard technologies such as the electrical power system (EPS), thermal control subsystem, communications and onboard computing (also referred to as command and data handling) systems, several specialized systems are required. This type of a mission may trigger particular attitude determination and control system (ADCS) needs, based on pointing accuracy requirements. Obviously, suitable imaging hardware is also required. Prior work [93] has described a prospective

1.4 Uses of Small Spacecraft

EPS. An ADCS solution is described in [94]. Example communications systems [95] and onboard computing systems [96] have also been previously presented. Several software technologies are also required for the mission. These include super-resolution, mosaicking, and task sharing between craft. Super-resolution enhances imagery beyond the physical collection capabilities of the craft. Mosaicking combines images together to produce a more ready-to-use data product and it eliminates the retransmission of overlapping areas. Task sharing between craft may be required to collect the level of data required to meet temporal and spatial coverage goals. Each of these technologies will now be discussed, followed by a discussion of an example collaborative mission.

1.4.1.1 Super-Resolution and Mosaicking

Super-resolution is used to enhance imagery; it produces a higher level of output resolution than the imagery fed to the engine. While a variety of single-source super-resolution algorithms exist (based on patterns in the image [97], heuristics [98, 99], and other techniques), these may place too much reliance on unsupported inference to be suitable for many applications. Multi-frame super-resolution algorithms (e.g., [100–102]) make use of subtle differences between images to make a more educated inference as to what the pixel configuration would be at a higher resolution. Super-resolution techniques may introduce false positives (nonexistent feature inclusion) and false negatives (feature exclusion) into the imagery [103]. It has been shown useful for processing raw imagery as well as other types of geospatial data [104, 105].

Mosaicking is used to combine multiple images into one composite one. The mosaicking software not only must identify the correct relative position of the two images (by lining up shared points, for example) but also may need to correct the shape of the images to match each other. Two common mosaicking techniques exist: Scale-Invariant Feature Transform (SIFT) and Speeded Up Robust Features (SURF) [106–108]. The use of mosaicking prepares the images to be directly useful to users, allowing possible transmission directly to the point of use (e.g., broadcasting to multiple handheld computers for use in the field), presuming that a sufficiently robust communication channel exists.

Previous work has considered the inclusion of super-resolution and mosaicking into small spacecraft missions [109, 110]. Ghosh et al. [111] have discussed the assessment of super-resolution and mosaicking performance in a CubeSat, albeit using high altitude balloon data. Finding more optimized algorithms (which may potentially make use of certain data not available in a general case scenario) will significantly enhance software and, prospectively, mission performance.

1.4.1.2 Task Sharing Between Craft

An orbital services model [112, 113] and various federated satellite service [114, 115] approaches have been proposed under which craft can collaborate to perform tasks. Under these models, supplier craft advertise the services that they can provide, and prospective consumers evaluate the service suppliers available to them, making a selection based on a combination of factors (e.g., quality, timeliness).

A CubeSat remote sensing mission could utilize an entire (or a subset thereof) network of providers; however, this same methodology could also be applied to a small cluster of craft which communicate and provide/consume services within the group.

Both image collection and processing could be distributed. This would facilitate greater temporal coverage (and the collection of the imagery required for multiple source super-resolution) as well as lower cost, by concentrating processing capabilities onto a subset of spacecraft.

1.4.1.3 Example Remote Sensing Mission

A collaborative small spacecraft mission could incorporate multiple spacecraft from a single operator or country, spacecraft from multiple operators/countries, or a combination of the foregoing. While an economic model could be devised entailing payments from one to another for services rendered, an alternate (perhaps easier to manage) approach would be to require a contribution proportionate to the level of benefit that is expected (or the attributable level of expense). This could range from some countries participating as partners in a spacecraft to others (who would enjoy more benefit) providing multiple craft.

Several considerations must be kept in mind as one is assessing the suitability of the data prospectively collected for a given application. First, of course, are the particulars of the application. Data that was suitable for one purpose within a general category (e.g., land cover assessment and planning) may not serve another (e.g., roadway planning). Thus, specific needs must be identified and the compliance of the solution with those needs assessed. Wertz et al. [74] discuss this process, both in general and in the context of an imaging spacecraft. Jensen [83] provides numerous examples of previous remote sensing missions and their capabilities.

The second consideration is the degradation of capabilities over the life of the mission or if one partner fails to deliver their equipment. In a cluster configuration, degradation can be gradual (as opposed to the all-or-nothing statuses provided by a single craft approach) with proper planning. For risk mitigation purposes, critical elements should be spread between partners and orbital locations. Through this, the impact of a partner failing to deliver, pulling out during operations, or equipment failure or damage can be mitigated. Special consideration may be required if a dominant partner is paired with numerous smaller ones in a given plan. The utility of the resulting cluster should be assessed under various possible degradation scenarios.

Third, the utility of a given resolution (or quality) of data should be considered relative to projected costs. Care should be taken to adjust costs (which, for example, for labor) may be commonly discussed in terms of the labor rates in more affluent regions. This should also be considered in the context of the build/buy decision.

1.4.2 Communications Mission

A communications mission would have similar requirements, with a few modifications. The imaging system would be largely unneeded (some sort of imaging system might be required if sun or star tracking was used as part of the ADCS) as would the supporting software technologies that were previously discussed. Alternately, it is likely that a more robust EPS would be required as (presumably) the communications spacecraft would operate continuously (relaying traffic, etc.). Additional generation capabilities may be required, likely in the form of additional solar panels. Some missions may use nonconventional approaches such as wireless power transmission [116] and more robust storage capabilities, as compared to generation capacity (for spacecraft whose use is sporadic). Obviously, a more robust communications system would be required. This might support different bands and full-duplex (concurrent send and receive) capabilities.

1.5 Conclusion

This chapter has provided an introduction to small spacecraft. It has briefly covered small satellites' origin, the origin of CubeSats, and how they can be utilized. It has introduced multiple classification schemes for small spacecraft, including those that group spacecraft by their size (mass), function, and whether they are used for university purposes or not. The benefits prospectively provided by small spacecraft missions and their development have been reviewed.

Through this, the chapter has laid a foundation for discussions in subsequent chapters. Chapter 2 extends the discussion of small spacecraft benefits and Chaps. 3 and 4 extend the discussion of their prospective uses. Later chapters discuss how to initiate a small spacecraft development program and assess it in an educational context.

References

1. Straub, J. 2012. Cubesats: A low-cost, very high-return space technology. In *Proceedings of the 2012 Reinventing Space Conference*. Los Angeles, CA.
2. — — —. 2013. Increasing national space engineering productivity and educational opportunities via intrepreneurship, entrepreneurship and innovation. *Technology and Innovation* 15: 211–226.
3. Straub, J., and D. Whalen. 2014. Evaluation of the educational impact of participation time in a small spacecraft development program. *Education Sciences* 4(1): 141–154.
4. Straub, J., J. Berk, A. Nervold, C. Korvald, and D. Torgerson. 2013. Application of collaborative autonomous control and the open prototype for educational NanoSats framework to enable orbital capabilities for developing nations. In *Proceedings of the 64th International Astronautical Congress*. Beijing, China.

5. Desell, T., R. Bergman, K. Goehner, R. Marsh, R. VanderClute, and S. Ellis-Felege. 2013. Wildlife@Home: Combining crowd sourcing and volunteer computing to analyze avian nesting video. In *2013 IEEE 9th International Conference on eScience (eScience)*.
6. De Selding, P.B. *Space Forecast Predicts Satellite Production Boom*. http://www.space.com/6839-space-forecast-predicts-satellite-production-boom.html.
7. Twiggs, R., and B. Malphrus. 2011. CubeSats. In *Space Mission Engineering: The New SMAD*, ed. J.R. Wertz, D.F. Everett, and J.J. Puschell, 803–821. Hawthorne, CA: Microcosm Press.
8. Bashevkin, E., J. Kenahan, B. Manning, B. Mahlstedt, and A. Kalman. 2012. A novel hemispherical anti-twist tracking system (HATTS) for CubeSats. In *Proceedings of the 26th AIAA/USU Conference on Small Satellites*.
9. Taraba, M., C. Rayburn, A. Tsuda, and C. MacGillivray. 2009. Boeing's CubeSat TestBed 1 attitude determination design and on-orbit experience. Presented at Proceedings of the AIAA/USU Conference on Small Satellites.
10. Padmanabhan, S., S. Brown, P. Kangaslahti, R. Cofield, D. Russell, R. Stachnik, J. Steinkraus, and B. Lim. 2013. A 6U CubeSat constellation for atmospheric temperature and humidity sounding. Proceeding of the AIAA/USU Conference on Small Satellites.
11. National Science Foundation. 2011. *CubeSat-based science missions for geospace and atmospheric research program solicitation* (Tech. Rep. NSF 12-536). National Science Foundation.
12. Woellert, K., P. Ehrenfreund, A.J. Ricco, and H. Hertzfeld. 2011. Cubesats: Cost-effective science and technology platforms for emerging and developing nations. *Advances in Space Research* 47(4): 663–684.
13. *DICE: Dynamic Ionosphere CubeSat Experiment*. http://www.sdl.usu.edu/downloads/dice.pdf.
14. Clark, C., K. Viergever, A. Vick, and I. Bryson. 2012. Achieving global awareness via advanced remote sensing techniques on 3U CubeSats.
15. Hill, T., J. Berk, J. Straub, J. Schiralli, B. Badders, and N.J. Long. 2013. Deep space planetary exploration using commercially available solar electric propulsion.
16. Klesh, A., J. Baker, J. Bellardo, J. Castillo-Rogez, J. Cutler, L. Halatek, E.G. Lightsey, N. Murphy, and C. Raymond. 2013. INSPIRE: Interplanetary NanoSpacecraft pathfinder in relevant environment. Presented at Proceedings of the AIAA/USU Small Satellite Conference. Logan, UT.
17. *CubeSat Personal Satellite Kit*. http://www.interorbital.com/CubeSat_1.htm.
18. *TubeSat Personal Satellite*. http://www.interorbital.com/TubeSat_1.htm.
19. *Pico-Class Products Price List*. http://www.tyvak.com/products/Pico/TyvakPriceList_v103_.pdf.
20. Berk, J., J. Straub, and D. Whalen. 2013. Open prototype for educational NanoSats: Fixing the other side of the small satellite cost equation. In *Proceedings of the 2013 IEEE Aerospace Conference*. Big Sky, MT.
21. *NASA Announces Candidates for CubeSat Space Missions*. http://www.nasa.gov/home/hqnews/2011/feb/HQ_11-038_CubeSat.html.
22. *SeeMe Industry Day Briefing*. http://www.darpa.mil/WorkArea/.
23. London, J., M. Ray, D. Weeks, and B. Marley. 2011. The first US army satellite in fifty years: SMDC-ONE first flight results.
24. Abramowitz L.R. Paul La Tour, Peter Mastro, Alan Frazier, Catherine Venturini, George Sondecker, and Lyle Abramowitz US air force's SMC/XR SENSE NanoSat program, AIAA SPACE 2011 Conference & Exposition.
25. *Project Description*. https://www.qb50.eu/project.php.
26. *First-Ever NSF Satellite Mission, Dynamic Ionosphere Cubesat Experiment, Launches Tomorrow at 5:48 AM ET*. http://twitter.com/#!/NSF/statuses/129637763745189889.
27. Fevig, R., J. Casler, J. Straub, R. Lilko, and C. Church. 2012. Blending research and teaching through small spacecraft development projects. In *2012 European Cubesat Symposium*. Brussels, Belgium.
28. Skrobot, G. 2012. ELaNa—"Making it happen!" In *The 9th Annual CubeSat Developers' Workshop*.
29. *Suppliers*. http://www.cubesat.org/index.php/collaborate/suppliers.

References

30. Milliron, R. 2012. Interorbital's N9 rocket: Test program, schedule, and launch manifest update. In *The 9th Annual CubeSat Developers' Workshop*.
31. *Frequently Asked Questions*. http://www.cubesatkit.com/content/faq.html.
32. *Intrepid Pico-Class CubeSat Suite*. http://www.tyvak.com/products/Pico/IntrepidSuite_201201b_.pdf.
33. Thakker, P., and W. Shiroma. 2010. *Emergence of Pico- and Nanosatellites for Atmospheric Research and Technology Testing*. Reston: AIAA Press.
34. 2012. Program. *1st Interplanetary CubeSat Workshop*.
35. Whalen, D.J. 2002. *The Origins of Satellite Communications, 1945–1965*. Washington, DC: Smithsonian Press.
36. *About Pumpkin, Inc.* http://www.cubesatkit.com/content/pumpkin/about_pumpkin_inc.html.
37. *History & Performance of Pumpkin's Products in Space*. http://www.cubesatkit.com/content/space.html.
38. Fitzsimmons, S., and A. Tsuda. 2013. Rapid development using tyvak's open source software model. Proceedings of the AIAA/USU Conference on Small Satellites.
39. *NRO Taps Boeing for Next Batch of Cubesats*. http://www.spacenews.com/military/100408-nro-taps-boeing-nextcubesats.html.
40. Campbell-Kelly, M., and W. Aspray. 2004. *Computer: A History of the Information Machine*, 2nd ed. Oxford: Westview Press.
41. Straub, J., and R. Fevig. 2012. Achieving educational outcomes through CubeSat curriculum incorporation. Proceedings of the 9th Annual Cubesat Workshop.
42. *MSU Satellite Orbits the Earth After Early Morning Launch*. http://www.montana.edu/cpa/news/nwview.php?article=10458.
43. *Explorer 1 Overview*. http://www.nasa.gov/mission_pages/explorer/explorer-overview.html.
44. Chirayath, V., and B. Mahlstedt. 2012. HiMARC 3D-high-speed, multispectral, adaptive resolution stereographic CubeSat imaging constellation. Presented at Proceedings of the AIAA/USU 2012 Small Satellite Conference.
45. Straub, J., and D. Whalen. 2013. An assessment of educational benefits from the OpenOrbiter space program. *Education Sciences* 3(3): 259–278.
46. Straub, J., D. Whalen, and R. Marsh. 2014. Assessing the value of the OpenOrbiter program's research experience for undergraduates. *Sage Open* 2014.
47. Taylor, M.W. 2007. Trashing the solar system one planet at a time: Earth's orbital debris problem. *Georgetown International Environmental Law Review* 20: 1–60.
48. Wertz, J.R. 2011. Space mission communities. In *Space Mission Engineering: The New SMAD*, ed. J.R. Wertz, D.F. Everett, and J.J. Puschell. Hawthorne, CA: Microcosm Press.
49. Larsen, J.A., and J.D. Nielsen. 2011. Development of CubeSats in an educational context. Presented at 2011 5th International Conference on Recent Advances in Space Technologies (RAST).
50. Alminde, L., M. Bisgaad, D. Vinter, T. Viscor, and K.Z. Østergard. 2004. The aaucubesat student satellite project: Architectural overview and lessons learned. Presented at 16th IFAC Symposium on Automatic Control in Aerospace,(Russia).
51. Bailey, J., S. Tsitas, D. Bayliss, and T. Bedding. 2012. A CubeSat mission for exoplanet transit detection and astroseismology. Presented at Proceedings of the 6U CubeSat Low Cost Space Missions Workshop.
52. Hand, E. 2015. Startup liftoff. *Science* 348(6231): 172–177. doi:10.1126/science.348.6231.172.
53. Weeks, D., A.B. Marley, and J. London III. 2009. SMDC-ONE: An army nanosatellite technology demonstration. Presented at SMDC-ONE: An Army Nanosatellite Technology Demonstration.
54. Cardin, J., K. Coste, D. Williamson, and P. Gloyer. 2003. A cold gas micro-propulsion system for CubeSats. Proceedings of the AIAA/USU Conference on Small Satellites.
55. Scharfe, M., K. Meinzer, and R. Zimmermann. 1996. Development of a magnetic-bearing momentum wheel for the AMSAT phase 3-D small satellite. Presented at International Symposium on Small Satellites.
56. Moreau, M.C., E.P. Davis, J.R. Carpenter, D. Kelbel, G.W. Davis, and P. Axelrad. 2002. Results from the GPS flight experiment on the high earth orbit AMSAT OSCAR-40 spacecraft. In *ION GPS-02 2002 Conference*.

57. Swartwout, M. 2004. University-class satellites: From marginal utility to 'disruptive' research platforms. Presented at 18th Annual AIAA/USU Conference on Small Satellites. 11pp.
58. ———. 1997. The role of universities in small satellite research. Presented at International Aerospace Conference. Moscow, Russia.
59. ———. 2013. The long-threatened flood of university-class spacecraft (and CubeSats) has come: Analyzing the numbers. Presented at Proceedings of the 27th Annual AIAA/USU Conference on Small Satellites.
60. ———. 2008. The first one hundred university-class spacecraft 1981–2008. *IEEE Aerospace and Electronic Systems Magazine* 24(3).
61. ———. 2015. Secondary spacecraft in 2015: Analyzing success and failure. In *Proceedings of the 2015 IEEE Aerospace Conference*. Big Sky, MT.
62. Deepak, R.A., and R.J. Twiggs. 2012. Thinking out of the box: Space science beyond the CubeSat. *Journal of Small Satellites* 1(1): 3–7.
63. Thakker, P., and G. Swenson. 2010. Survey of atmospheric and other research projects employing small university satellites. In *Emergence of Pico- and Nanosatellites for Atmospheric Research and Technology Testing*, ed. P. Thakker and W. Shiroma, 63–67. Reston, VA: American Institute of Aeronautics and Astronautics.
64. Swartwout, M. 2012. A statistical survey of rideshares (and attack of the CubeSats, part deux). Presented at 2012 IEEE Aerospace Conference.
65. Thakker, P., and G. Swenson. 2010. Management and implementation of a CubeSat interdisciplinary senior design course. In *Emergence of Pico- and Nanosatellites for Atmospheric Research and Technology Testing*, ed. P. Thakker and W. Shiroma. Reston: AIAA.
66. Zhou, C. 2012. Teaching engineering students creativity: A review of applied strategies. *Journal of Efficiency and Responsibility in Education and Science* 5(2): 99–114.
67. Crawley, E., J. Malmqvist, S. Östlund, and D. Brodeur. 2007. *Rethinking Engineering Education—The CDIO Approach*. New York: Springer.
68. Smith, M.W., D.W. Miller, and S. Seager. 2011. Enhancing undergraduate education in aerospace engineering and planetary sciences at MIT through the development of a CubeSat mission. Presented at SPIE Optical Engineering Applications.
69. Akagi, J., T. Tamashiro, W. Tonaki, and W. Shiroma. 2010. Small satellites 101: University of Hawaii small-satellite program. In *Emergence of Pico- and Nanosatellites for Atmospheric Research and Technology Testing*, ed. P. Thakker and W. Shiroma, 33–60. Reston, VA: American Institute of Aeronautics and Astronautics.
70. Straub, J., R. Fevig, J. Casler, and O. Yadav. 2013. Risk analysis & management in student-centered spacecraft development projects. In *Proceedings of the 2013 Reliability and Maintainability Symposium*. Orlando, FL.
71. Straub, J., R. Marsh, and D. Whalen. 2015. The impact of an interdisciplinary space program on computer science student learning. *Journal of Computers in Mathematics and Science Teaching* 34: 97–125.
72. Woellert, K. 2012. Space access: Still the major issue for the small satellite community. *Journal of Small Satellites* 1: 45–47.
73. Skrobot, G., and R. Coelho. 2012. ELaNa–Educational launch of nanosatellite: Providing routine RideShare opportunities. Presented at Proceedings of SmallSat Conference.
74. Wertz, J.R., D.F. Everett, and J.J. Puschell. 2011. *Space Mission Engineering: The New SMAD*. Hawthorne, CA: Microcosm Press.
75. Swartwout, M. 2011. AC 2011-1151: Significance of student-built spacecraft design programs it's impact on spacecraft engineering education over the last ten years. Presented at Proceedings of the American Society for Engineering Education Annual Conference. http://www.asee.org/file_server/papers/attachment/file/0001/1307/paper-final.pdf.
76. ———. 2013. Cheaper by the dozen: The avalanche of rideshares in the 21st century. Presented at 2013 IEEE Aerospace Conference.
77. Bergsrud, C., and J. Straub. 2013. A 6-U commercial constellation for space solar power supply to other spacecraft. Presented at Spring 2013 CubeSat Workshop.

78. Noca, M., F. Jordan, N. Steiner, T. Choueiri, F. George, G. Roethlisberger, N. Scheidegger, H. Peter-Contesse, M. Borgeaud, and R. Krpoun. 2009. Lessons learned from the first swiss pico-satellite: SwissCube. Proceedings of the AIAA/USU Conference on Small Satellites.
79. Hunyadi, G., J. Ganley, A. Peffer, and M. Kumashiro. 2004. The university nanosat program: An adaptable, responsive and realistic capability demonstration vehicle. Presented at IEEE Aerospace Conference Proceedings, 2004.
80. 13 February 2013. *Call for Proposals: Fly Your Satellite!* http://www.esa.int/Education/Call_for_Proposals_Fly_Your_Satellite.
81. Garvey, J., and E. Besnard. 2004. Development status of a nanosat launch vehicle. In *AIAA Paper (2004-4065)*.
82. Milliron, R. 2013. Interorbital's NEPTUNE dedicated SmallSat launcher: 2013 test milestones and launch manifest update. In *2013 Spring CubeSat Developers' Workshop*. San Luis Obispo, CA.
83. Jensen, J.R. 2009. *Remote Sensing of the Environment: An Earth Resource Perspective*, 2nd ed. Upper Saddle River: Prentice Hall.
84. Kalluri, S., P. Gilruth, and R. Bergman. 2003. The potential of remote sensing data for decision makers at the state, local and tribal level: Experiences from NASA's synergy program. *Environmental Science & Policy* 6(6): 487–500.
85. Seelan, S., D. Baumgartner, G. Casady, V. Nangia, and G. Seielstad. 2007. Empowering farmers with remote sensing knowledge: A success story from the US Upper Midwest. *Geocarto International* 22(2): 141–157.
86. Seelan, S.K., S. Laguette, G.M. Casady, and G.A. Seielstad. 2003. Remote sensing applications for precision agriculture: A learning community approach. *Remote Sensing of Environment* 88(1): 157–169.
87. Kim, H.J., D.R. Olsen, and S. Laguette. 2012. International space station agricultural camera (ISSAC) sensor onboard the international space station (ISS) and its potential use on the earth observation. Presented at ASPRS 2012 Annual Conference.
88. Weng, Q. 2012. Remote sensing of impervious surfaces in the urban areas: Requirements, methods, and trends. *Remote Sensing of Environment* 117: 34–49.
89. Miller, R.B., and C. Small. 2003. Cities from space: Potential applications of remote sensing in urban environmental research and policy. *Environmental Science & Policy* 6(2): 129–137.
90. Tralli, D.M., R.G. Blom, V. Zlotnicki, A. Donnellan, and D.L. Evans. 2005. Satellite remote sensing of earthquake, volcano, flood, landslide and coastal inundation hazards. *ISPRS Journal of Photogrammetry and Remote Sensing* 59(4): 185–198.
91. Rajitha, K., C. Mukherjee, and R. Vinu Chandran. 2007. Applications of remote sensing and GIS for sustainable management of shrimp culture in India. *Aquacultural Engineering* 36(1): 1–17.
92. Chen, Q., Y. Zhang, A. Ekroos, and M. Hallikainen. 2004. The role of remote sensing technology in the EU water framework directive (WFD). *Environmental Science & Policy* 7(4): 267–276.
93. Chaieb, S., M. Wegerson, J. Straub, R. Marsh, B. Kading, and D. Whalen. 2015. The OpenOrbiter CubeSat as a system-of-systems (SoS). Presented at 10th International Conference on System of Systems Engineering.
94. Wegerson, M., M. Partridge, N. Crocker, D. Schindele, B. Friend, L. Lewis, B. Johnson, J. Straub, and R. Marsh. 2015. Hardware design for an intelligent attitude determination and control system (ADCS). Presented at University of North Dakota School of Graduate Studies Scholarly Forum.
95. Wegerson, M., J. Straub, and R. Marsh. 2015. A low-cost radio for an OPEN CubeSat. Presented at University of North Dakota School of Graduate Studies Scholarly Forum.
96. ———. 2015. Design of an onboard distributed multiprocessing system for a CubeSat. Presented at University of North Dakota School of Graduate Studies Scholarly Forum.
97. Freeman, W.T., T.R. Jones, and E.C. Pasztor. 2002. Example-based super-resolution. *IEEE Computer Graphics and Applications* 22(2): 56–65.

98. Trifas, M., and J. Straub. 2011. Super resolution: A database driven inference approach. Presented at Proceedings of the 15th World Multi-Conference on Systemics, Cybernetics and Informatics.
99. Straub, J. 2012. Robotic applications of a defensible error-aware super-resolution technique. In *The Seventeenth International Symposium on Artificial Life and Robotics 2012*.
100. Tsai, R., and T.S. Huang. 1984. Multiframe image restoration and registration. *Advances in Computer Vision and Image Processing* 1(2): 317–339.
101. Borman, S., and R.L. Stevenson. 1998. Super-resolution from image sequences-a review. Presented at Proceedings of 1998 Midwest Symposium on Circuits and Systems, 1998.
102. Elad, M., and A. Feuer. 1999. Super-resolution reconstruction of image sequences. *IEEE Transactions on Pattern Analysis and Machine Intelligence* 21(9): 817–834.
103. Trifas, M., and J. Straub. 2012. A comparison of techniques for super-resolution evaluation. Presented at IS&T/SPIE Electronic Imaging.
104. Straub, J. 2013. Difference modeling enhancement of topographic super-resolution. Presented at SPIE Defense, Security, and Sensing.
105. –––. 2012. Super-resolution terrain map enhancement for navigation based on satellite imagery. Presented at IS&T/SPIE Electronic Imaging.
106. Juan, L., and O. Gwun. 2009. A comparison of SIFT, PCA-SIFT and SURF. *International Journal of Image Processing (IJIP)* 3(4): 143–152.
107. Lowe, D.G. 2004. Distinctive image features from scale-invariant keypoints. *International Journal of Computer Vision* 60(2): 91–110.
108. Bay, H., T. Tuytelaars, and L. Van Gool. 2006. Surf: Speeded up robust features. In *Computer Vision–ECCV*, ed. A. Leonardis, H. Bischof, and A. Pinz. Heidelberg: Springer.
109. Bhatia, A., K. Goehner, J. Sand, J. Straub, A. Mohammad, C. Korvald, and A.K. Nervold. 2013. Sensor and computing resource management for a small satellite. Presented at IEEE Aerospace Conference, 2013.
110. Straub, J., and R.A. Fevig. 2012. Earth impactors: Threat analysis and multistage intervention mission architecture. Presented at SPIE Defense, Security, and Sensing.
111. Ghosh, D., N. Kaabouch, and R.A. Fevig. 2014. Robust spatial-domain based super-resolution mosaicing of CubeSat video frames: Algorithm and evaluation. *Computer and Information Science* 7(2): 68.
112. Straub, J., A. Mohammad, J. Berk, and A.K. Nervold. 2013. Above the cloud computing: Applying cloud computing principles to create an orbital services model. Presented at SPIE Defense, Security, and Sensing.
113. Straub, J. 2013. Spatial computing in an orbital environment: An exploration of the unique constraints of this special case to other spatial computing environments. Presented at Proceedings of the 2013 Spatial Computing Workshop at the Autonomous Agents and Multi-Agent Systems (AAMAS) 2013 Conference.
114. Golkar, A. 2013. Architecting federated satellite systems for successful commercial implementation. Presented at AIAA Space.
115. Grogan, P.T., S. Shirasaka, A. Golkar, and O.L. de Weck. 2014. Multi-stakeholder interactive simulation for federated satellite systems. Proceedings of the 2014 IEEE Aerospace Conference.
116. Bergsrud, C., and J. Straub. 2014. A space-to-space microwave wireless power transmission experiential mission using small satellites. *Acta Astronautica* 103: 193–203.

Chapter 2
Why Start a Small Spacecraft Program

This chapter focuses on the reasons behind starting a small spacecraft program. In this context, both programs started for educational benefit and those launched for research or other purposes are considered. An overview of reasons for program initiation is provided. Then, research and educational benefits are discussed. Broader benefits that could be provided by a small spacecraft program to society-at-large are then considered. Finally, the chapter discusses the potential of using a small spacecraft program to develop, demonstrate, or advance national space competency, before concluding.

2.1 Overview

Small spacecraft development activity is increasing significantly. Between 2000 and 2013, the number of manifested "university-class" spacecraft has increased from below 5 to over 35 [6]. In 2014, just under 30 university-class spacecraft were launched [7]. From its initial design by Jordi Puig-Suari and Robert Twiggs in 2000 [8], the CubeSat standard (one type of small spacecraft that is gaining in popularity due to its easy-to-integrate common form factor [9]) has matured from a tool for student learning to a mechanism for conducting bona fide science [10, 11] and other work [12]. Interest in CubeSats has been buoyed by low-cost [13] and free-to-qualified developer launch services, available through NASA's Educational Launch of Nanosatellites program [14] and the ESA [15]. Interest is also being generated

---

This chapter is based on, revises and extends the papers "Evaluation of the Educational Impact of Participation Time in a Small Spacecraft Development Program" [1], "Student Expectations from Participating in a Small Spacecraft Development Program" [2], "An Assessment of Educational Benefits from the OpenOrbiter Space Program" [3], "OpenOrbiter: A Low-Cost, Educational Prototype CubeSat Mission Architecture" [4] and "Application of Collaborative Autonomous Control and the OPEN Prototype for Educational NanoSats Framework to Enable Orbital Capabilities for Developing Nations" [5].

for larger-sized small spacecraft. In Europe, the ESA's Student Space Exploration and Technology Initiative [16] has generated larger spacecraft (similar to the size and mass to spacecraft facilitated in the United States by the Air Force's University NanoSat Program [17]). The European Student Earth Orbiter, for example, is a 45-kg spacecraft with dimensions of approximately 30 cm × 30 cm × 100 cm [18]. While still being built by universities [6], these spacecraft are being constructed by government [19] and industry [19]. Small spacecraft are now even being considered for lunar [20] and interplanetary use [21].

While the benefits of the form factor for missions are clear, the reasons for student involvement in the design and development of a small spacecraft are less so. In many cases, students participate and devote their skills to small spacecraft development on a voluntary basis (or at a wage level below what they could make by obtaining an off-campus job). Do these students seek to work in the space engineering field? What reasons drive those students who are studying ancillary topics? These questions are considered in the following sections that begin the process of assessing why students decide to participate in small spacecraft development and what benefits they hope to obtain from doing so.

2.2 Research Benefits

First, the reasons for why small spacecraft are developed are considered. Swartwout [22] proffers that the role of the "university-class" spacecraft (a type of spacecraft with education as its primary objective and increased risk tolerance because of the academic environment) is to provide an opportunity to try things that could not be effectively explored on larger, more expensive missions due to risk management and other concerns. Many have also used them to provide the educational experience for students envisioned by Puig-Suari and Twiggs when initially defining the CubeSat form factor [23]. As of 2014, nearly 100 educational institutions have developed a small spacecraft (some in collaboration with other institutions) and several have developed more than one craft [6]. The use of the CubeSat form factor has expanded beyond academia: over 50 CubeSats not originating from an academic institution are manifested for launch in 2013 (compared to only 30 from academia) [6]. Academic institutions are also involved in the development of a limited number of non-CubeSat-class spacecraft.

Despite what the foregoing might suggest, the development of small spacecraft isn't new. Some would argue that small spacecraft have their foundations in the earliest launches. Sputnik is pointed to, by some, as an example of a small satellite. Dickson, for example, describes it as being the "size of a beach ball" and weighing "a mere 184 lb" [24]. Thinking of something this size as small is not unsurprising, considering the size of many current and historical spacecraft. Intelsat 10, a communications satellite launched in 2004, had an initial launch mass of 5600 kg [25], for example.

It would be another 40 years, however, until the event that would drive their phenomenal growth. In 2000, Bob Twiggs (then at Stanford leading the Satellite Quick

2.2 Research Benefits

Research Testbed project) challenged the notion of the size of a small satellite [23]. The Orbiting Picosatellite Automatic Launcher (OPAL) deployed six "hockey puck-sized" spacecraft, weighing 1 kg [23]. Following this success, Twiggs and Jordi Puig-Suari developed specifications for the CubeSat form factor (see, e.g., [26]) and developed the commonly used launcher: the Poly-PicoSatellite Orbital Deployed (P-POD) [27]. The first CubeSat was launched in 2003. To date, more than 200 CubeSats have successfully reached orbit [7, 28]; numerous others have been developed and lost to launch failures or never launched [29]. Twiggs is not stopping with the CubeSat form factor he is now working on making small satellites even smaller by developing a form factor for a satellite one-eighth the volume of a CubeSat (5 cm×5 cm×5 cm) called the PocketQub [23]. This spacecraft is targeted at enhancing high school STEM education.

Are small spacecraft just educational tools then? Thakker and Swenson [30] suggest that this may be the case. They contend that "most university satellite programs have focused more on their educational missions" than on advancing science and developing new techniques for science and engineering. Several examples of science missions exist, however, including the University of Illinois ION-1 (oxygen airglow photometer) and ION-2 (neutral hydrogen photometer) spacecraft and Taylor University's TEST (Langmuir plasma probe, electric field boom, VLF receiver, SSD spectrometer, and transient photometer) and TU SAT-1 (Langmuir plasma probe, tether, and Nitol tether). Swartwout, however, disagreed he proffered [22], in 2004, that "university-class satellites" could be "disruptive" research platforms: they can alter the way that space research is carried out. He asserted that this disruptive capability comes from the particular strengths of research universities: students' enthusiasm and novel ideas and the "freedom to fail" [22].

In disagreement with Thakker and Swenson's statement only 2 years earlier, in 2012, Swartwout [31] noted that university programs have moved away from being "beepsats" (a term used to characterize spacecraft lacking "a compelling science, technology, or communications payload") to incorporating real scientific, engineering, or other goals. These missions, he noted (in 1997), should have risk from their unique characteristics and not be an exercise in navigating complexity [32]. A university program, under these circumstances, can be beneficial to students' educational attainment and investigate "risky and/or innovative methods" [32].

From an educational perspective, university missions are valuable to industry and others as they can and do employ the same mission analysis and design techniques [8, 33, 34] utilized by industry, military and government, preparing students for workforce entry. Chin et al. [35] proffer, however, that the standardization in the CubeSat development community is critical to the form factor's success; this of course is atypical for space missions which (while reusing proven/qualified components) generally implement mission or program-specific designs. The notion of a more standardized approach to space missions ties in with a proposed TRL 10 paradigm, where operations of a model of spacecraft are characterized and failure conditions are well known and understood (as is, for example, typical of commercial airliners) [36].

Small spacecraft are not just valuable for educational activities, education research and training future researchers. Small spacecraft are also being used to perform bona fide research. CubeSats, for example, are pushing technical boundaries. Twiggs and Malphrus [8] provide an overview: CubeSats are using (and in some cases being used to test) advances such as plastic printed structures, deployable solar panels and technologies (such as Stanford's Hemispherical Anti-Twist Tracking System, HATTS [37]), advanced propulsion (e.g., heated Freon gas), and 3D printed propulsion.

Given the foregoing, it might seem that small spacecraft are excellent tools for both research and education. Swartwout [38], however, highlights two key problems: spacecraft projects are not responsive to university needs of creating a sustained educational program or attracting external research sponsorship. Prior to the advent of CubeSats, Swartwout proffers that schools "rarely, if ever" completed a second project after an initial success. CubeSats, he asserts, are changing this; however, it is unclear as to whether this has changed significantly, except for at a few key schools.

2.3 Educational Benefits

Educational benefits from small spacecraft come from both formal and informal learning. Formal learning occurs in lectures, readings, and other structured activities in courses and elsewhere. Informal learning comes in the form of project-based learning (PBL). PBL is a technique where students learn by doing. While the concept is by no means new (as the apprenticeship style of learning has been used throughout history [39, 40]), it is seen as a departure from the traditional lecture-based style of instruction. The benefits of PBL are seen by some as so great as to have an effect on national competitiveness on an international scale. Gilmore [41], for example, contends that STEM education will determine the future of nations and proffers that PBL and EE are critical to the United States' ability to compete globally.

2.3.1 *Experiential Learning and Problem-Based Learning*

Project-based learning (also known as problem-based learning or experiential learning) involves providing students with a challenge to solve or a problem to resolve. Students collect information, assess the nature of the challenge or problem, and devise and implement a plan to achieve the assigned goal or resolve the assigned problem. The utility of PBL techniques has been demonstrated for all stages of education ranging from primary to university level (see [42–47]). The use of PBL has also been favorably assessed in numerous disciplines, such as computer science [48, 49], computer engineering [50], electrical engineering [51, 52], mechanical

2.3 Educational Benefits

engineering [53–55], aerospace engineering [56, 57], management [58], project management [59], and entrepreneurship [60] and marketing [61]. Small spacecraft development, in an educational setting, is inherently an exercise in PBL. Students can be involved (depending on program particulars) in the design, development, testing, and operations of the spacecraft. PBL small spacecraft programs (e.g., [47, 62]) have been shown to be effective in achieving educational outcomes.

The development of small spacecraft and CubeSats provides students with PBL style educational benefits [47, 63, 64] in their discipline of participation. From the foregoing it is clear that PBL is effective in a diverse number of disciplines relevant to small spacecraft development. It has also been shown to be effective across a wide range of educational and age levels [43, 47].

Student small spacecraft development provides participants with the opportunity to develop and hone their skills. Students will also inherently develop new 'out-of-the-box' concepts. The educational environment allows them to try these concepts and to make mistakes, on a path to success in a low-risk environment facilitated by the low mission cost levels [64].

In a college or university context, PBL can occur in several formats. Students may engage in PBL activities as part of a regular course, such as a course project [47] and a PBL course. They may participate as part of an independent or directed study [3] or to satisfy a senior design or capstone requirement [65]. They may also participate for extracurricular educational enrichment [3]. Small satellites easily integrate into a project-based learning (PBL) methodology. The PBL technique seeks to create student learning through immersion in a project. Students are tasked with overcoming foreseen and unforeseen challenges and learn during the process.

PBL has also been shown to deliver benefits in addition to driving learning about course topics. These include improved student self-image [66], creativity [67], motivation [66], material understanding [68], workforce preparation [68], job placement [69], and academic program [70] and knowledge retention [71]. Zhou [72] contends that creativity is critical for engineers. This creativity can be developed via a variety of techniques including creating a conducive environment requiring problem solving. Zhou identifies PBL as a technique that can help create engineering creativity through student-centered, self-directed collaborative exercises. To this end, an eight-step approach is proposed beginning with (1) "problem setting," incorporating, (2) brainstorming, (3) systematization, (4) thematic selection, (5) formulation of learning tasks, (6) knowledge acquisition via self-studying, (7) knowledge integration, and concluding with (8) structuring the knowledge in terms of the problem at hand.

Smith et al. [73] demonstrated a technique specifically for incorporating CubeSat development in undergraduate aerospace engineering and planetary science curriculum. Their approach is based on prior work by Crawley et al. [74] who pioneered an approach entitled "Conceive-Design-Implement-Operate" (CDIO), based upon feedback from numerous engineering education stakeholders (educators, industry, students, etc.). Smith et al. [73] expand this by asserting that there is a significant need, in aerospace engineering, for shared understanding between scientists and engineers. In the ExoplanetSat initiative, students from the Department of Earth,

Atmospheric and Planetary Sciences were involved in the design process, via enrolment in the three-semester CDIO course progression. This required students to engage in a science versus engineering trade process throughout the mission, analogous to how a larger mission of this type would be performed. While this expanded the scope of interdisciplinary collaboration slightly, it still did not fully encompass all discipline types that would be required to be involved in a real-world mission of this type.

Another small satellite PBL example is provided by Rodriguez-Osorio and Ramirez [75] who presented work at the ETSI de Telecomunicación in Madrid, Spain, related to an extracurricular NanoSat project. This 21-month project was student conceived and implemented (under faculty supervision). An antenna array designed for the purpose of inter-spacecraft communication (for a CubeSat-size craft) was created and its performance characterized. Rodriguez-Osorio and Ramirez proffer that this experiment demonstrates the feasibility of implementing simulated industry-analog engineering projects with limited resources and "promising results" [20].

Prior work has demonstrated the efficacy of small spacecraft development, for student learning, in general [1, 3]. It has also considered benefits that were specific to undergraduates [76], computer science students [77], and various roles within the development group [1, 3]. More details on and an expansion of this prior work are discussed in Chaps. 8–10.

While student-involved projects may provide significant benefits (whether attempting to achieve exclusively learning goals or a combination of substantive research and education), it is important to note that they carry significant risk of project failure or less-than-complete success [78, 79]. This risk comes from conventional risk sources (e.g., delays beyond project manager control, supplier issues); many elements of conventional risk are also exacerbated by the project conditions typical of student projects (e.g., participant lack of knowledge and inexperience). Student-involved projects also incorporate their own particular risk factors driven by the academic environment (e.g., a prioritization of course work over project performance, students joining and leaving the project at semester breaks and other times). Risk and general management are, thus, crucially important. Risk, mitigation, and management are discussed further in Chap. 7.

2.3.2 Benefits of Interdisciplinary Projects

Interdisciplinary projects are typical of the modern workplace. Most undertakings of any size cannot be performed exclusively by practitioners of a single discipline or specialty. However, many student projects in an academic environment are performed within the context of a course or a degree program. Because of this, they generally involve a set of similarly trained students working on a narrowly defined topic. Even projects that span disciplines (e.g., teams participating in NASA's Lunabotics competition [80]) may be limited to only closely related disciplines (electrical, mechanical, and computer engineering, for example).

Because of this, students may not gain exposure to a true interdisciplinary project (characterized by multiple specialists collaboratively performing work related to their area of specialty) until after they enter the workforce. This may require them to unlearn practices and approaches learned while working only in discipline-constrained teams. They may also experience frustration if the process of getting up to speed in this impairs their performance during their initial period (normally including some sort of an evaluation/probation process) with a new employer whom they are trying to impress.

Involving students in interdisciplinary work prevents 'silo'-type work habits from developing; students instead learn how to work well in collaboration with others with skills divergent from their own. In addition to these general benefits, students also begin to learn the particular vernacular and work styles of the disciplines whose practitioners-in-training they collaborate with. Interdisciplinary projects may also be able to have a larger scale than those within a single discipline, offering an opportunity for project management practices and discipline-specific multiperson collaboration techniques (e.g., software version control management) to be learned and refined. All of this increases student participant preparation for workplace entry and success.

2.4 Societal Benefits

This section considers the societal benefits that can be provided by small satellites and could, consequentially, be produced by a small spacecraft program. To this end, it begins with a discussion of the benefits produced by remote sensing, a common use for satellites: the data products that can be produced and their prospective uses are considered. A brief discussion of the required technologies to produce these benefits is then provided (this was discussed in greater detail in Chap. 1). Then, a discussion of one particular mission concept that may offer particular benefit for developing countries is presented. This is followed by a discussion of the qualitative assessment of spacecraft data and a discussion of the prospective role small spacecraft can have in developing national space competency.

2.4.1 Remote Sensing Benefits, Data Products, and Their Uses

While a wide variety of data products can be generated via remote sensing (including thermal infrared imaging, multispectral imaging, microwave and LIDAR imaging, and gravitational data [81]), one of the most common is visible light imagery (and near-visible light imagery that can be produced via changing the filtering applied to standard sensing equipment).

Visible light remote sensing data is defined by its coverage and spatial and temporal resolution as well as other qualitative aspects [81]. Coverage of desired areas is clearly important. Spatial resolution is a measure of the size represented by each pixel on the imagery. The utility of spatial resolution levels ranging from submeter to over a kilometer has been demonstrated [82]. Temporal resolution is a measure of how frequently data for a given area can be reobtained (i.e., how current the data is).

For agriculture, visible light sensing data can be useful for assessing where to deploy and the deployment of fungicides and pesticides, assessing crop damage due to weather and assessing drainage patterns and designing drainage solutions [83]. The utility of both aerial and satellite imagery has been demonstrated for this purpose [81, 83, 84]. The resolution required varies by application; however, the utility of 20-m data has been demonstrated by the International Space Station Agricultural Camera [85]; the use of 10-m [83] and much higher resolution [84] data has also been demonstrated. Temporal requirements vary, based on the desired phenomena under study. In addition, data may be needed on demand for use in storm damage assessment and other cases.

Remote sensing of urban areas, such as might be used for municipal planning and other purposes, has been performed with data ranging from 100-m to 10-m and higher resolution [86]. This data can be used to determine material composition, land cover, and land use for planning and other purposes. The level of temporal coverage required varies significantly, with once-a-year resolution being acceptable for some applications and more frequent imagery (including imagery at given times) being required for others. Miller and Small [87] proffer that remote sensed imagery is particularly important for developing regions as it can serve to replace the growth and environmental condition data collected in situ (or by other means) in more developed nations. They also note the utility of remotely sensed data being unobstructive and consistent.

Remote sensed data has also been shown to be useful in responding to hazards such as earthquakes, volcanos, floods, landslides, and costal inundations. In this context, data can be used to prioritize response efforts, direct responders as well as to, in the longer term, perform risk assessments, and establish policies to prevent future issues [88].

A multitude of other uses for remote-sensed data exist, many of which would be relevant to both developed and developing nations. These include its use in aquaculture (sea farming), such as was demonstrated in India [89], and water policy [90].

2.4.2 Technologies and Mission

This section briefly highlights the technologies needed for remote sensing missions. It includes both required technologies and those with augmentative capabilities. Multiple technologies are required to support a prototypical remote sensing mission for developing countries. These include basic systems (such as attitude determination and control, the electric power system, communications, and thermal control) as well as mission-specific technologies. Some software technologies are also

2.4 Societal Benefits

required to enhance mission performance: these include super-resolution, mosaicking, and task sharing between craft. Super-resolution enhances imagery beyond the physical collection capabilities of the craft. Mosaicking combines images together to produce a more ready-to-use data product and it eliminates the retransmission of overlapping areas. Task sharing between craft may be required to collect the level of data required to meet temporal and spatial coverage goals. An expanded discussion of this topic was presented in Chap. 1.

2.4.3 Collaborative Mission for Developing Countries

A collaborative mission is one prospective way that smaller countries could afford access to space as well as the sensing capabilities that they require. Such a collaborative mission could incorporate multiple spacecraft from a single country, spacecraft from multiple countries, or a combination of the foregoing. While an economic model could be devised entailing payments from one to another for services rendered, an alternate (perhaps easier to manage) approach would be to require a contribution proportionate to the level of benefit that is expected (or the attributable level of expense). This could range from some countries participating as partners in a spacecraft to others (who would enjoy more benefit) providing multiple craft.

There is no requirement that the collaborating countries be neighbors; in fact, collaboration between dispersed countries may be ideal as this may reduce contention for craft use when over a group of local collaborators' general vicinity. It would, of course, be necessary for all of the target regions to be able to be served from the selected orbits. However, the use of other resources (e.g., processing capabilities) would require only sufficient communications opportunities to exist between orbital craft.

2.4.4 Qualitative Analysis

A discussion of how to assess the utility of a particular mission approach is now presented. The quality and utility of the data and the suitability for various applications are considered.

2.4.4.1 Quality and Utility of Data

The quality and utility of remote-sensed data will be a function of several different factors. First is the quality of the collection equipment and supporting subsystems that aim, stabilize, and point it. Even if the data is of a suitably high resolution, if it is blurry or otherwise degraded, it may be of little use. Image processing techniques may be able to resolve (or mitigate the impact of) some imperfections.

Second, the number of images that can be collected of the given region in a given period of time will directly affect the prospective uses that it is suitable for. If multiple images can be captured of a given area in a support period of time, computational image enhancement [91] can be performed. The number of images that can be collected constrains the level of enhancement that will be possible.

2.4.4.2 Application Suitability

The spatial resolution of the enhanced data will be a function of the collection hardware, selected orbit, and the level of software enhancement (if any). Several considerations must be kept in mind as one is assessing the suitability of the data prospectively collected for a given application. First, of course, are the particulars of the application. Data that was suitable for one purpose within a general category (e.g., land cover assessment and planning) may not serve another (e.g., roadway planning). Thus, specific needs must be identified and the compliance of the solution with those needs assessed. Wertz et al. [8] discuss this process, both in general and in the context of an imaging spacecraft. Jensen [81] provides numerous examples of previous remote sensing missions and their capabilities.

The second consideration is the degradation of capabilities over the life of the mission (or if one partner fails to deliver their equipment for a shared cluster mission). In a cluster configuration, degradation can be gradual (as opposed to the all-or-nothing statuses provided by a single craft approach) with proper planning. In clusters, for risk mitigation purposes, critical elements should be spread between partners and orbital locations. Through this, the impact of a partner failing to deliver, pulling out during operations, or equipment failure or damage can be mitigated.

Third, the utility of a given resolution (or quality) of data should be considered relative to projected costs. Alternate prospective suppliers and/or other data collection techniques should also be considered.

2.4.4.3 Mission Approach Considerations

All of the different prospective mission approaches have benefits and drawbacks. The collaborative mission, for example, requires cooperation between countries. This may be difficult to attain and maintain. Single craft missions have a significantly increased risk due to the craft (and its critical components) representing single points of failure.

Export control regimes and various political considerations should be considered. They may limit who can work on a mission and/or impede or preclude the collaborative mission pproach. As the product of fundamental research, some university-developed spacecraft (such as the current OPEN design) may enjoy favorable treatment under US export regulations. Modified versions (created by commercial entities) or commercially developed designs may not enjoy this benefit.

Additionally, an analog to the fundamental research classification may not exist or be as favorable under other nations' laws.

Finally, technical problems, a lack of skilled staff, training issues, and other hurdles may impede deployment. This possibility should be considered in build/buy decisions. Risk is discussed further in Chap. 7.

2.5 Considerations Based on National Space Competency

Some countries will undoubtedly see small spacecraft development programs as a mechanism to increase their national space competency. In these cases, the desire to develop 'home grown' technology (or to avoid technology where continued access may be subject to the continued friendship with another nation or which is regulated by another nation's export control regime) may trump many other considerations. Chapter 3 presents a discussion of one of the key choices faced by developers in starting a program: whether to build from scratch, buy a vendor kit, or use a hybrid of the two choices. The reason for program initiation (i.e., if it is for national competency building) may, obviously, be a key factor in this type of decision.

2.6 Conclusion

This chapter has presented a discussion of why various entities and individuals decide to start (or participate in) a small spacecraft development program. It has discussed the prospective benefits from technical development, research, and education perspectives. It has also, briefly, considered the role of the program (in a national context) on this decision-making process.

References

1. Straub, J., and D. Whalen. 2014. Evaluation of the educational impact of participation time in a small spacecraft development program. *Education Sciences* 4(1): 141–154.
2. ———. 2013. Student expectations from participating in a small spacecraft development program. *Aerospace* 1(1): 18–30.
3. ———. 2013. An assessment of educational benefits from the OpenOrbiter space program. *Education Sciences* 3(3): 259–278.
4. Straub, J., C. Korvald, A. Nervold, A. Mohammad, N. Root, N. Long, and D. Torgerson. 2013. OpenOrbiter: A low-cost, educational prototype CubeSat mission architecture. *Machines* 1: 1–32.
5. Straub, J., J. Berk, A. Nervold, C. Korvald, and D. Torgerson. 2013. Application of collaborative autonomous control and the open prototype for educational NanoSats framework to enable orbital capabilities for developing nations. In *Proceedings of the 64th International Astronautical Congress*, Beijing, China.

6. Swartwout, M. 2013. The long-threatened flood of university-class spacecraft (and CubeSats) has come: Analyzing the numbers. Presented at Proceedings of the 27th Annual AIAA/USU Conference on Small Satellites.
7. ———. 2015. Secondary spacecraft in 2015: Analyzing success and failure. In *Proceedings of the 2015 IEEE Aerospace Conference*. Big Sky, MT
8. Wertz, J.R., D.F. Everett, and J.J. Puschell. 2011. *Space Mission Engineering: The New SMAD*. Hawthorne, CA: Microcosm Press.
9. California Polytechnic State University. 2009. *CubeSat design specification, revision 12*. California Polytechnic State University, San Luis Obispo, CA, 1 Aug 2009.
10. Muylaert, J., R. Reinhard, C. Asma, J. Buchlin, P. Rambaud, and M. Vetrano. 2009. QB50: An international network of 50 CubeSats for multi-point, in-situ measurements in the lower thermosphere and for re-entry research. Presented at ESA Atmospheric Science Conference. Barcelona, Spain.
11. Chirayath, V., and B. Mahlstedt. 2012. HiMARC 3D-high-speed, multispectral, adaptive resolution stereographic CubeSat imaging constellation. Presented at Proceedings of the AIAA/USU 2012 Small Satellite Conference.
12. Weeks, D., A.B. Marley, and J. London III. 2009. SMDC-ONE: An army nanosatellite technology demonstration. Presented at SMDC-ONE: An Army Nanosatellite Technology Demonstration.
13. Garvey, J., and E. Besnard. 2004. Development status of a nanosat launch vehicle. In *AIAA Paper (2004-4065)*.
14. Skrobot, G., and R. Coelho. 2012. ELaNa–Educational launch of nanosatellite: Providing routine RideShare opportunities. Presented at Proc. SmallSat Conference.
15. 13 February 2013. *Call for Proposals: Fly Your Satellite!* http://www.esa.int/Education/Call_for_Proposals_Fly_Your_Satellite.
16. Arana, L. 2003. SSETI and ESEO-launching the dream! *ESA Bulletin* 116: 70–73.
17. Hunyadi, G., J. Ganley, A. Peffer, and M. Kumashiro. 2004. The university nanosat program: An adaptable, responsive and realistic capability demonstration vehicle. Presented at 2004 IEEE Aerospace Conference Proceedings, 2004.
18. Bruzzi, D., P. Tortora, F. Giulietti, and P. Galeone. 2013. European student earth orbiter: ESA's educational microsatellite program. Proceedings of the AIAA/USU Conference on Small Satellites.
19. Norton, C.D., M.P. Pasciuto, P. Pingree, S. Chien, and D. Rider. 2012. Spaceborne flight validation of NASA ESTO technologies. Presented at 2012 IEEE International Geoscience and Remote Sensing Symposium (IGARSS).
20. Walker, R., and M. Cross. 2010. The European student moon orbiter (ESMO): A lunar mission for education, outreach and science. *Acta Astronautica* 66(7): 1177–1188.
21. Staehle, R., D. Blaney, H. Hemmati, M. Lo, P. Mouroulis, P. Pingree, T. Wilson, J. Puig-Suari, A. Williams, and B. Betts. 2011. Interplanetary CubeSats: Opening the solar system to a broad community at lower cost. Presented at CubeSat Developers' Workshop. Logan, UT.
22. Swartwout, M. 2004. University-class satellites: From marginal utility to 'disruptive' research platforms. Presented at 18th Annual AIAA/USU Conference on Small Satellites. 11pp.
23. Deepak, R.A., and R.J. Twiggs. 2012. Thinking out of the box: Space science beyond the CubeSat. *Journal of Small Satellites* 1(1): 3–7.
24. Dickson, P. 2001. *Sputnik: The Shock of the Century*. New York, NY: Walker Publishing Company, Inc.
25. Martin, D., P. Anderson, and L. Bartamian. 2007. *Communications Satellites*, 5th ed. El Segundo, CA: Aerospace Press.
26. California Polytechnic State University. 2013. *CubeSat design specification, revision 13*. California Polytechnic State University, San Luis Obispo, CA, 19 Aug 2013.
27. Puig-Suari, J., C. Turner, and W. Ahlgren. 2001 Development of the standard CubeSat deployer and a CubeSat class PicoSatellite. Presented at 2001 IEEE Aerospace Conference Proceedings.
28. Stephenson, H. 2011. The next big thing is small. *ASK Magazine* 38.

29. Swartwout, M. 2008. The first one hundred university-class spacecraft 1981–2008. *IEEE Aerospace and Electronic Systems Magazine* 24(3): A1–A23.
30. Thakker, P., and G. Swenson. 2010. Survey of atmospheric and other research projects employing small university satellites. In *Emergence of Pico- and Nanosatellites for Atmospheric Research and Technology Testing*, ed. P. Thakker and W. Shiroma, 63–67. Reston, VA: American Institute of Aeronautics and Astronautics.
31. Swartwout, M. 2012. A statistical survey of rideshares (and attack of the CubeSats, part deux). Presented at 2012 IEEE Aerospace Conference.
32. ———. 1997. The role of universities in small satellite research. Presented at International Aerospace Conference. Moscow, Russia.
33. Larson, W.J., and J.R. Wertz. 1999. *Space Mission Analysis and Design*, 3rd ed. Hawthorne, CA: Microcosm Press.
34. Fortescue, P., G. Swinerd, and J. Stark. 2011. *Spacecraft Systems Engineering*, 4th ed. Chichester: Wiley.
35. Chin, A., R. Coelho, L. Brooks, R. Bugent, and J. Puig-Suari. 2008. Standardization promotes flexibility: A review of CubeSats' success. In *Proceedings of the 6th Responsive Space Conference*. Los Angeles, CA.
36. Straub, J. 2015. In search of technology readiness level (TRL) 10. *Aerospace Science and Technology* 46: 312–320.
37. Bashevkin, E., J. Kenahan, B. Manning, B. Mahlstedt, and A. Kalman. 2012. A novel hemispherical anti-twist tracking system (HATTS) for CubeSats. In *Proceedings of the 26th AIAA/USU Conference on Small Satellites*.
38. Swartwout, M. 2006. Bandit: A platform for responsive educational and research activities. In *Proceedings of the 4th Responsive Space Conference*. Los Angeles, CA
39. Snell, K.D. 1996. The apprenticeship system in British history: The fragmentation of a cultural institution. *History of Education* 25(4): 303–321.
40. Elbaum, B. 1989. Why apprenticeship persisted in Britain but not in the United States. *Journal of Economic History* 49(2): 337–349.
41. Gilmore, M. 2013. Improvement of STEM education: Experiential learning is the key. *Modern Chemistry & Applications* 1: e109.
42. Mountrakis, G., and D. Triantakonstantis. 2012. Inquiry-based learning in remote sensing: A space balloon educational experiment. *Journal of Geography in Higher Education* 36(3): 385–401.
43. Mathers, N., A. Goktogen, J. Rankin, and M. Anderson. 2012. Robotic mission to mars: Hands-on, minds-on, web-based learning. *Acta Astronautica* 80: 124–131.
44. Fevig, R., J. Casler, and J. Straub. 2012. Blending research and teaching through near-earth asteroid resource assessment. Presented at Space Resources Roundtable and Planetary & Terrestrial Mining Sciences Symposium.
45. Hall, S.R., I. Waitz, D.R. Brodeur, D.H. Soderholm, and R. Nasr. 2002. Adoption of active learning in a lecture-based engineering class. Presented at Proceedings of the 32nd Annual Frontiers in Education Conference.
46. Brodeur, D.R., P.W. Young, and K.B. Blair. 2002. Problem-based learning in aerospace engineering education. Presented at Proceedings of the 2002 American Society for Engineering Education Annual Conference and Exposition.
47. Straub, J., J. Berk, A. Nervold, and D. Whalen. 2013. OpenOrbiter: An interdisciplinary, student run space program. *Advances in Education* 2: 4–10.
48. Correll, N., R. Wing, and D. Coleman. 2013. A one-year introductory robotics curriculum for computer science upperclassmen. *IEEE Transactions on Education* 56(1): 54–60. doi:10.1109/TE.2012.2220774.
49. Broman, D., K. Sandahl, and M. Abu Baker. 2012. The company approach to software engineering project courses. *IEEE Transactions on Education* 55(4): 445–452. doi:10.1109/TE.2012.2187208.
50. Qidwai, U. 2011. Fun to learn: Project-based learning in robotics for computer engineers. *ACM Inroads* 2(1): 42–45.

51. Bütün, E. 2005. Teaching genetic algorithms in electrical engineering education: A problem-based learning approach. *International Journal of Electrical Engineering Education* 42(3): 223–233.
52. de-Camargo-Ribeiro, L.R. 2008. Electrical engineering students evaluate problem-based learning (PBL). *International Journal of Electrical Engineering Education* 45(2): 152–161.
53. Robson, N., I.S. Dalmis, and V. Trenev. 2012. Discovery learning in mechanical engineering design: Case-based learning or learning by exploring? Presented at 2012 ASEE Annual Conference.
54. Coller, B.D., and M.J. Scott. 2009. Effectiveness of using a video game to teach a course in mechanical engineering. *Computers & Education* 53(3): 900–912.
55. Das, S., S.A. Yost, and M. Krishnan. 2010. A 10-year mechatronics curriculum development initiative: Relevance, content, and results—Part I. *IEEE Transactions on Education* 53(2): 194–201.
56. Saunders-Smits, G.N., P. Roling, V. Brügemann, N. Timmer, and J. Melkert. 2012. Using the engineering design cycle to develop integrated project based learning in aerospace engineering. Presented at International Conference on Innovation, Practice and Research in Engineering Education.
57. Jayaram, S., L. Boyer, J. George, K. Ravindra, and K. Mitchell. 2010. Project-based introduction to aerospace engineering course: A model rocket. *Acta Astronautica* 66(9): 1525–1533.
58. Reynolds, M., and R. Vince. 2004. Critical management education and action-based learning: Synergies and contradictions. *Academy of Management Learning & Education* 3(4): 442–456.
59. Pollard, C.E. 2012. Lessons learned from client projects in an undergraduate project management course. *Journal of Information Systems Education* 23(3): 271–282.
60. Okudan, G.E., and S.E. Rzasa. 2006. A project-based approach to entrepreneurial leadership education. *Technovation* 26(2): 195–210.
61. Siegel, C.F. 2000. Introducing marketing students to business intelligence using project-based learning on the World Wide Web. *Journal of Marketing Education* 22(2): 90–98.
62. Nielsen, J.F.D., X. Du, and A. Kolmos. 2010. Innovative application of a new PBL model to interdisciplinary and intercultural projects. *International Journal of Electrical Engineering Education* 47(2): 174–188.
63. Larsen, J.A., and J.D. Nielsen. 2011. Development of cubesats in an educational context. Presented at 2011 5th International Conference on Recent Advances in Space Technologies (RAST).
64. Larsen, J.A., J.F.D. Nielsen, and C. Zhou. 2013. Motivating students to develop satellites in problem and project-based learning (PBL) environment. *International Journal of Engineering Pedagogy* 3(3): 11–17.
65. Thakker, P., and W. Shiroma. 2010. *Emergence of Pico- and Nanosatellites for Atmospheric Research and Technology Testing*. Reston: AIAA Press.
66. Doppelt, Y. 2003. Implementation and assessment of project-based learning in a flexible environment. *International Journal of Technology and Design Education* 13(3): 255–272.
67. Ayob, A., R.A. Majid, A. Hussain, and M.M. Mustaffa. 2012. Creativity enhancement through experiential learning. *Advances in Natural and Applied Science* 6(2): 94–99.
68. Simons, L., L. Fehr, N. Blank, H. Connell, D. Georganas, D. Fernandez, and V. Peterson. 2012. Lessons learned from experiential learning: What do students learn from a practicum/internship? *International Journal of Teaching and Learning in Higher Education* 24(3): 325–334.
69. Breiter, D., C. Cargill, and S. Fried-Kline. 2013. An industry view of experiential learning. *Hospitality Review* 13(1): 8.
70. Edwards, A., S.M. Jones, E. Wapstra, and A.M. Richardson. 2012. Engaging students through authentic research experiences. Presented at Proceedings of the Australian Conference on Science and Mathematics Education (Formerly UniServe Science Conference).
71. Bauerle, T.L., and T.D. Park. 2012. Experiential learning enhances student knowledge retention in the plant sciences. *HortTechnology* 22(5): 715–718.
72. Zhou, C. 2012. Teaching engineering students creativity: A review of applied strategies. *Journal of Efficiency and Responsibility in Education and Science* 5(2): 99–114.

73. Smith, M.W., D.W. Miller, and S. Seager. 2011. Enhancing undergraduate education in aerospace engineering and planetary sciences at MIT through the development of a CubeSat mission. Presented at SPIE Optical Engineering Applications.
74. Crawley, E., J. Malmqvist, S. Östlund, and D. Brodeur. 2007. *Rethinking Engineering Education—The CDIO Approach*. New York: Springer.
75. Rodriguez-Osorio, R.M., and E.F. Ramírez. 2012. A hands-on education project: Antenna design for inter-CubeSat communications [education column]. *IEEE Antennas and Propagation Magazine* 54(5): 211–224.
76. Straub, J., D. Whalen, and R. Marsh. 2014. Assessing the value of the OpenOrbiter program's research experience for undergraduates. *Sage Open* 2014.
77. Straub, J., R. Marsh, and D. Whalen. 2015. The impact of an interdisciplinary space program on computer science student learning. *Journal of Computers in Mathematics and Science Teaching* 34: 97–125.
78. Straub, J., R. Fevig, J. Casler, and O. Yadav. 2013. Risk analysis & management in student-centered spacecraft development projects. In *Proceedings of the 2013 Reliability and Maintainability Symposium*. Orlando, FL.
79. Straub, J. 2014. Extending the student qualitative undertaking involvement risk model. *Journal of Aerospace Technology and Management* 6(3): 333–352.
80. Mueller, R.P. 2011. Lunabotics mining competition: Inspiration through accomplishment. Presented at Earth and Space 2012@ Struction, and Operations in Challenging Environments.
81. Jensen, J.R. 2009. *Remote Sensing of the Environment: An Earth Resource Perspective*, 2nd ed. Upper Saddle River: Prentice Hall.
82. Kalluri, S., P. Gilruth, and R. Bergman. 2003. The potential of remote sensing data for decision makers at the state, local and tribal level: Experiences from NASA's synergy program. *Environmental Science & Policy* 6(6): 487–500.
83. Seelan, S., D. Baumgartner, G. Casady, V. Nangia, and G. Seielstad. 2007. Empowering farmers with remote sensing knowledge: A success story from the US upper midwest. *Geocarto International* 22(2): 141–157.
84. Seelan, S.K., S. Laguette, G.M. Casady, and G.A. Seielstad. 2003. Remote sensing applications for precision agriculture: A learning community approach. *Remote Sensing of Environment* 88(1): 157–169.
85. Kim, H.J., D.R. Olsen, and S. Laguette. 2012. International space station agricultural camera (ISSAC) sensor onboard the international space station (ISS) and its potential use on the earth observation. Presented at ASPRS 2012 Annual Conference.
86. Weng, Q. 2012. Remote sensing of impervious surfaces in the urban areas: Requirements, methods, and trends. *Remote Sensing of Environment* 117: 34–49.
87. Miller, R.B., and C. Small. 2003. Cities from space: Potential applications of remote sensing in urban environmental research and policy. *Environmental Science & Policy* 6(2): 129–137.
88. Tralli, D.M., R.G. Blom, V. Zlotnicki, A. Donnellan, and D.L. Evans. 2005. Satellite remote sensing of earthquake, volcano, flood, landslide and coastal inundation hazards. *ISPRS Journal of Photogrammetry and Remote Sensing* 59(4): 185–198.
89. Rajitha, K., C. Mukherjee, and R. Vinu Chandran. 2007. Applications of remote sensing and GIS for sustainable management of shrimp culture in India. *Aquacultural Engineering* 36(1): 1–17.
90. Chen, Q., Y. Zhang, A. Ekroos, and M. Hallikainen. 2004. The role of remote sensing technology in the EU water framework directive (WFD). *Environmental Science & Policy* 7(4): 267–276.
91. Freeman, W.T., T.R. Jones, and E.C. Pasztor. 2002. Example-based super-resolution. *IEEE Computer Graphics and Applications* 22(2): 56–65.

Chapter 3
To Build, Buy, or in Between?

This chapter covers the process of defining a small spacecraft program. It is not intended to replace or replicate robust models, such as those discussed by Wertz et al. [6] and Fortescue and Swinerd [7]. Nor is it meant to provide a light-weight approximation (such as was presented in [8]). Instead, this chapter seeks to help the reader evaluate and answer key questions regarding spacecraft program formulation.

This begins with a discussion about why firms, institutions, and governments build and launch small spacecraft. Then it begins the exploration of the prospective approaches with a discussion of the benefits and drawbacks of a kit-based approach, where most parts are procured from a vendor and integrated by the developers, with mission-specific payload components. Next the bespoke (built-from-scratch) approach is discussed. Then, the notion of developing based on a preexisting framework is considered. Finally, hybrid approaches and a decision-making framework are discussed, before concluding.

3.1 Why Launch a NanoSat?

Building a tightly coupled miniaturized satellite can be a challenging proposition even for experienced developers. This challenge is even greater for first time researchers and students. Some seek to demonstrate their capability to develop a

---

This chapter is based on, revises, and extends the papers "OpenOrbiter: A Low-Cost, Educational Prototype CubeSat Mission Architecture" [1], *"Evaluation of the Educational Impact of Participation Time in a Small Spacecraft Development Program"* [2], *"Increasing National Space Engineering Productivity and Educational Opportunities via Intrepreneurship, Entrepreneurship and Innovation"* [3], *"CubeSats: A Low-Cost, Very High-Return Space Technology"* [4], *and "The Open Prototype for Educational NanoSats: Fixing the Other Side of the Small Satellite Cost Equation"* [5].

small spacecraft. Others, though, just want access to space for commercial, engineering, or scientific purposes. These groups can make meaningful contributions to the advancement of space science and technology if empowered with the right tools. This section will discuss why developers seek to develop and launch small spacecraft, informing later discussions of how to best accomplish this goal.

3.1.1 Low-Cost Test Platform

Small spacecraft can provide a low-cost, low-risk test platforms for sensors, actuators, propulsion systems, and other technology test and demonstration needs. By substantially reducing the costs associated with a technology failure (including one that results in spacecraft lost), higher-risk technologies can be tested leading to greater innovation and enhanced results.

3.1.2 Capability to Mature Technical Readiness of Experimental Space Technologies

Small spacecraft can, using existing launch infrastructure, be used to buy down mission risk at a lower price than many other approaches, advancing technology readiness levels (TRLs).

In aerospace development TRLs range from 1 to 9 (TRL level 10 has been proposed [9]) as a relative measure of a system's maturity. Each number corresponds to a specific level of development progress [10]. Due to funding source constraints and other factors gaps exist that result in some promising space technologies not being advanced to the next level of readiness.

CubeSats and other small spacecraft may remove barriers for integrating experimental technologies into a free flying space platform by allowing low-cost testing in a relevant operational environment (a key metric for TRL advancement). Experimental technologies that can be miniaturized or are already small enough to fit in a CubeSat can be tested in space relatively inexpensively. In this way, both technologists and federal funding agencies are able to get a greater return on their R&D spending through a test platform that can verify the characteristics and performance of mid-level TRL projects.

Commercial CubeSat kits are presently available to use for TRL advancement; however, a bespoke or framework approach may offer several advantages, particularly in applications where multiple spacecraft are required. Because a spacecraft design can be reproduced in quantity, without recurring the Research Development Test and Evaluation (RDT&E) expenses built into commercial kit pricing, these approaches may be advantageous for those experimenting with satellite constellations or scientific investigations requiring multiple spacecraft. A need to alter or integrate with vendor kits may also create interoperability or performance issues, especially if vendors do not supply low level hardware and software documentation. Bespoke, framework and hybrid approaches do not suffer from this issue. However, they also do not benefit from the vendor's knowledge, experience, testing, and hardware heritage.

3.1.3 Ecosystem of Innovation

A low-cost, low-risk, and versatile development platform drives experimentation with innovative and advanced technologies. The use of framework approaches and vendor kits encourages new development by freeing CubeSat programs to focus their resources on payloads and targeted subsystem innovation. The propensity for customization on framework-based approach satellites may increase the level and scope of experimentation and thus expand the depth and breadth of capabilities possible in small spacecraft form factors. As new technologies are successfully demonstrated an ecosystem of innovation may emerge.

Popular mobile phone manufacturers have leveraged a similar development paradigm (called crowdsourced innovation, by some). A community of software developers has evolved organically and created new software products for use on low-cost, easily obtainable hardware. Millions of new and first time software developers have emerged, generating a vibrant software marketplace [11].

Not unlike the mobile phone 'app' ecosystem, low-cost CubeSats and other small spacecraft may provide a development platform that is scalable from low budget first-time space operations to programs performing complex technology and science investigations. Framework and kit users can initiate a space program without prior experience working with spaceflight hardware, though expertise in the relevant engineering domains is of course required. Small spacecraft thus present an opportunity to capture new spaceflight developers who may be attracted by the low barrier to entry and the possibility of doing something new in outer space.

3.2 Overview of Different Approaches to Spacecraft Development

The space age began in the late 1950s as a contest to orbit the world's first artificial satellites. Sputnik 1 and Explorer 1, the first Soviet and American satellites, each had a mass of less than 100 kg [12, 13]. In the years that followed, the average mass of satellite systems has increased dramatically with the average satellite mass, during 2003–2011, being over 4000 kg [14]. These large and expensive spacecraft are critical to national security, weather forecasting, and communications. Their development requires a large base of skilled developers.

Recognizing this trend and the need to educate future aerospace engineers, Bob Twiggs and Jordi Puig-Suari collaborated to develop a standardized 10 cm × 10 cm × 10 cm satellite form factor commonly referred to as a CubeSat [15]. CubeSats provide students with a hands-on opportunity to work with and operate actual spacecraft hardware. The skills developed while working with CubeSat hardware are applicable to larger spacecraft development projects, such as those prevalent in industry and government. CubeSats also provide an opportunity for testing innovative new technologies that are not at a mature technological readiness level suitable for incorporation in a higher-risk, higher-cost mission [16, 17]. For these reasons, CubeSats have become a popular educational and research satellite platform. The small size of CubeSats generally has meant lower production costs, shorter development timelines, and less project complexity.

Because of their small size and mass, CubeSats have significantly lower launch costs than larger spacecraft. The cost paid per kilogram on a converted Soviet ICBM Dnepr rocket, for example, is $3000 [18]. Emerging launch providers appear poised to offer a greater number of low-cost, competitive launch solutions. CubeSat launch costs may, in the near future, be as low as $10,000 for a 1-kg spacecraft [4]. Alternately, government programs (such as NASA's ELaNa) make low- or no-cost launches available, on a secondary space available basis, to qualified institutions.

With launch costs for small spacecraft affordable, cost considerations now turn to the development of the spacecraft itself. Several techniques can be used. With a build-from-scratch approach, initial development costs approximately $250,000, based on in-house design, development, fabrication and testing using a combination of paid and volunteer student labor [4]. Not all missions need to incur this expense, however. The entry of commercial venders has reduced the cost of a single mission significantly. A robust kit, currently available from the Tyvak Nano-Satellite Systems Company for about $42,000, contains all required supporting subsystems [19] and requires development, integration and testing of the payload elements. These kits are, for some users, cost-effective on a single mission basis; however, they tend to increase costs over multiple missions due to recurring payments for vender-incurred research, development, testing, and engineering (RDT&E) expenses above and beyond the actual value of the hardware purchased.

Framework-based approaches may offer the best of both worlds. They reduce single mission costs (as the development has already been done by the framework provided) while also offering the lower recurring costs of a bespoke design and approach the bespoke level of adaptability. Each of the aforementioned approaches, as well as techniques to mix-and-match approach elements are now discussed.

3.3 Kit-Based Approach

The first small satellites can be likened to the Altair 8800. They demonstrated the technical capabilities and little more. However, the current generation of small satellites is poised to perform real science and engineering demonstrations that focus on advancing the state of the art (in addition to various CubeSat bus technologies and subsystems).

So how close are CubeSats to being the analog of the personal computer? There is no doubt that that they are far beyond the means of most individuals. However, the kit-based approach is the closest analog to the modern use-out-of-the-box personal computer. Even with a $50,000 price point [20, 21] for a fully functional 1-U CubeSat and launch services (which neglects the costs of actually building the payload components), there is little chance of the United States becoming a nation of satellite owners.

Small satellites, and kits in particular, provide several pathways towards greater public participation in space. First, they facilitate the training of students who will become the next generation of spacecraft engineering professionals—for both small and large spacecraft [22]. The lower-cost, lower-risk platform allows students to take lead roles in a way that simply wouldn't be possible on more risk-adverse missions. The diminished risk aversion also facilitates missions that try out new technologies, concepts, and even new paradigms.

Second, the lower cost facilitates the entry of small and mid-sized businesses into lead project roles and as satellite owners. These businesses lack the capitalization to build a multimillion dollar satellite; however, a satellite with a $50–$100 thousand dollar price may be achievable.

Third, small satellites facilitate the use of orbital remote sensing for smaller-scale research projects. This allows a project investigator to choose the best dates and times to collect the most meaningful data (e.g., matching the dates that in situ validation may be scheduled for—or when an important phenomena is expected to occur) as opposed to being limited to commercially produced data. This control also allows the scientist to task the satellite to focus on a not-previously scheduled phenomena of interest, should one be discovered during the research.

Finally, by making small satellites more accessible to the aforementioned—and others—a user/owner base is being formed. Based on the history of most electronic devices, this proliferation should result in even lower prices, making satellite ownership and access available to even more institutions. The formation of a negative cost spiral and positive ownership/access spiral is possible and highly desirable.

A key enabler of the small satellite proliferation is miniaturization. A recent mission which, in some ways, duplicated a much earlier mission provides an antidotal example of this. In 2011, Montana State University launched the Explorer One Prime Satellite [23]. This 1-U CubeSat was designed to "replicate the scientific mission" of the Explorer One spacecraft, launched in 1958, which detected the Van Allen Radiation Belt [23]. The 1958 model weighed 14 kg and had a dedicated launch, the 2011 satellite weighed less than 1.33 kg and shared a ride with other craft [23, 24]. If kits reach a production level where amortized development costs are spread over thousands of units, they will likely be the approach-of-choice for missions not focusing on hardware development and which don't require significant customization.

3.4 Bespoke Approach

CubeSats have largely been developed on a bespoke basis. Their small size has lowered production costs and reduced development timelines and project complexity. These qualities, in combination, make an excellent educational, science, and technology development platform attractive to many users. The National Aeronautics and Space Administration (NASA), Department of Defence (DoD), and commercial and educational institutions have all experimented with CubeSats [25]. While a large satellite may incur launch expenses (as a primary payload), on average, of $99 million, CubeSats face a dramatically lower launch cost. Prior work has discussed that the bespoke design approach has initial development costs of approximately $250,000, based on in-house design, development, fabrication, and testing using a combination of paid and volunteer student labor [4]. However, bespoke designs of subsequent craft (even without reuse) may be significantly lower for experienced developers. The bespoke approach offers the greatest flexibility and may be the only viable approach for missions that have significantly different form factors and/or hardware from previous missions.

3.5 Framework-Based Approach

In defining the OPEN framework, exemplary benefits were defined which include (1) lowering the cost of entry, for educational institutions, to operating a small spacecraft program, (2) lowering research program cost via allowing project-specific modifications to subsystems without requiring vendor negotiation or redesign from scratch, (3) facilitating student involvement in real research, instead of education-centric integration-only projects, and (4) facilitating initial mission efforts becoming a program by allowing initial mission performance to directly translate to program performance (and not requiring a substantial design cost outlay between integrating a vendor-provided kit spacecraft and producing a locally designed spacecraft affordable to teaching budgets).

With OPEN, a CubeSat can be developed for a parts cost (excluding payload-specific components) of approximately $5000 [26]. This is significantly less than the $40,000 or more that might be spent buying a one-time-use kit-based spacecraft or the $250,000 cost of developing the designs from scratch [4]. These lower cost levels may facilitate greater penetration of spacecraft development or spacecraft-based experiments into the educational systems of more affluent countries and enable spacecraft development in less-affluent ones [27]. Other framework approaches may offer a similar set of benefits.

3.6 Qualitative Evaluation of the Value of the Approaches

This section considers the costs and benefits of the foregoing approaches. For each category, the costs and benefits of approaches are discussed and quantified (to the extent possible). Factors that may contribute to the level of the cost / benefit received

3.6 Qualitative Evaluation of the Value of the Approaches   43

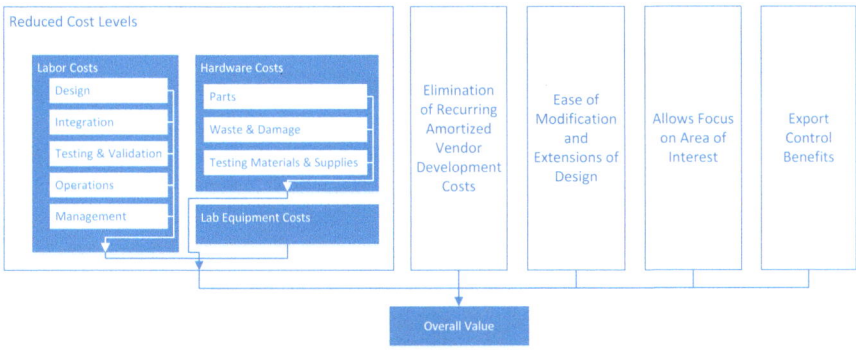

Fig. 3.1 Model for assessing value of spacecraft approach

are discussed. A value model is presented in Fig. 3.1 and expanded upon in the following sections.

3.6.1 Cost Levels

The reduced cost levels made possible by a framework such as OPEN stem from two sources: the freely available designs that reduce the amount of design and development work that is required and the costs that are reduced by design simplifications and optimizations.

Twiggs and Malphrus [25] proffer that the design and implementation of a CubeSat-class spacecraft, from scratch, and its testing may cost as much as $250,000. The exact amount of budget attributed to activities that are eliminated or reduced is, of course, different for each project. As has already been discussed, a kit approach may cost in the $50,000 range, though an approach costing only a fraction of this was presented, as an example, in [5].

For a framework approach, projects that utilize the base subsystems without modifications will incur the lowest level of expense: the design, development, integration, and testing of their payload and the fabrication, integration, and testing of the base subsystems. Those that elect to make significant modifications or replace base subsystems with significantly redesigned ones will, of course, incur higher development and testing costs in these areas.

The impact of using the framework designs on overall cost merits significant analysis, as this may be a driving reason to use (or not use) the design for many developers. There are two key components to the cost model (labor costs and hardware costs), presented in Fig. 3.2, which are directly influenced by the decision to use a framework, to develop from scratch, or to procure a vendor kit. An additional area (lab equipment costs) may be impacted, depending on the developer's current equipment holdings.

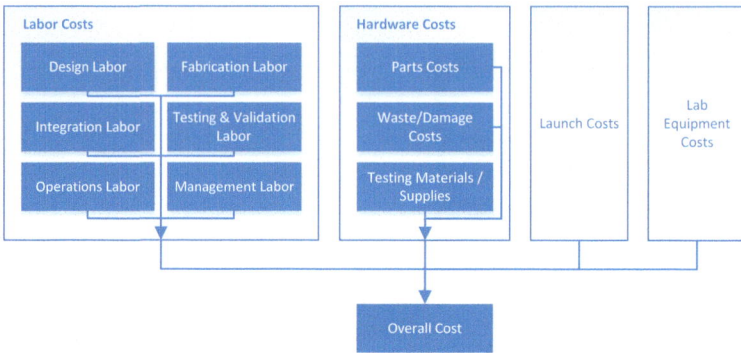

Fig. 3.2 Model for cost of spacecraft

3.6.1.1 Labor Costs

Labor costs can be decomposed into six categories: design labor, fabrication labor, integration labor, testing and validation labor, operations labor, and management labor. Each will now be briefly considered.

Design Labor

Relative to the build-from-scratch approach, the use of a framework will significantly reduce the need for design labor. The amount of savings enjoyed will depend on how significantly the developer modifies the framework designs, with those making no or limited modifications enjoying greater levels of savings (compared to those making more extensive modifications). Relative to the vendor kit approach, the design labor requirements for using the framework design without modification of the core components should be similar. In both instances, the developer still incurs some level of design costs related to designing their payload elements (and any necessary interface hardware, etc.). A framework may simplify this as it provides an example payload module that can be used as a guide or starting point for design. Some vendors provide payload boards or starter kits, which may provide a similar benefit.

Integration Labor

Compared to the design-from-scratch approach, the required level of integration labor for a framework-based spacecraft should be roughly equivalent. In both cases, all of the fabricated components must be assembled and their combined operation tested. The vendor kit approach should have a lower level of integration

3.6 Qualitative Evaluation of the Value of the Approaches   45

required, as most vendors pretest components to ensure that they will function when assembled. The framework approach may outperform the design-from-scratch approach in that the prior testing of the designs and their space qualification may reduce the number of times that boards must be refabricated or repaired. This reduces the need to reperform testing to validate the performance of the repaired or refabricated boards.

Testing and Validation Labor

The testing and validation labor requirements for the framework design should be significantly less than build-from-scratch approach and greater than the vendor kit-based approach. The build-from-scratch approach will require more testing and validation labor than the framework approach because of the necessity of testing hardware for which the design and implementation have not validated previously. Both the framework and build-from-scratch approaches will require board/subsystem implementation validation. The framework approach's implementation validation should require less time because the test plan (for the base framework system) can be supplied. This will need to be developed for the build-from-scratch approach. For the vendor kit approach, this validation is performed by the vendor prior to shipment. Because of the possibility of damage in shipment, a limited level of validation must still be performed; however, this can focus on testing processes instead of lower level testing designed to facilitate the identification of error sources (if a vendor part fails higher level testing, generally the most desirable course of action is to return the part to the vendor for replacement, instead of trying to diagnose the problem). Developers of all three approaches will need to devote time to integration and system level validation.

Developers who use the base framework system (or make minor modifications) will require less testing and validation labor than those who modify it significantly. The vendor kit approach, presuming that testing can be limited to verifying that no shipping damage has occurred and verifying integration success, should require the least labor in this area. The unmodified/minimally modified framework approach, significantly modified framework approach, and build-from-scratch approach will require (in order) progressively more labor.

Operations Labor

The level of operations labor that is required should not be significantly different between the three approaches. Any particular implementation may, of course, include features that impact operations labor required, however.

Management Labor

Each approach requires different management techniques and management involvement in a different set of areas. Labor is required for managing the performance of work that is done; it is also required to manage the procurement process for items that are sourced from a vendor.

The vendor kit approach requires significant management involvement in the vendor selection process. The choice of vendor and specific product will impact virtually every aspect of the mission, including the ease of payload integration and whether the final craft is able to meet mission requirements. This process is largely a management one (perhaps with nonmanagers performing research to support the decision-making process). In addition to vendor and product selection, management labor will be required to supervise the payload development, integration, testing and validation, and mission operations areas.

The build-from-scratch approach will require management oversight of all phases; however, there are no phases that are particularly management intensive (that is, mostly performed by managers as opposed to staff). These phases include design, development, integration, testing, and mission operations. It is expected that the integration and testing phases will take longer and thus require more management time than with the vendor kit approach. Additionally, the design and development (excluding payload development) management time will be in addition to the time requirements of the vendor kit approach. It is anticipated that for virtually all missions the vendor selection management will be significantly less than the management required for the build-from-scratch approach.

The framework approach requires management in all phases, like the build-from-scratch approach and (similarly) no one phase is particularly management intensive. The design phase will be limited to making any modifications required to the framework designs and payload design. Thus, less work is required during this phase and this, thus, means a commensurate reduction in the number of management hours that must be spent. Reductions, compared to the build-from scratch approach, will also be enjoyed during the development/fabrication, integration, and testing/validation phases, due to the fact that the core framework design may already have been validated, limiting the amount of design-attributable errors that must be detected and rectified. It is expected that the level of management labor required for a framework design will be more than is required for a vendor kit and less than is required for using the build-from-scratch approach.

Summary

The framework approach, overall, will require more labor than the vendor kit approach and less labor than the design-from-scratch approach. Table 3.1 presents a comparison of the level of labor required in the different areas discussed.

3.6 Qualitative Evaluation of the Value of the Approaches

Table 3.1 Comparison of the level of labor required

	OPEN	From-scratch	Vendor kit
Design	Low	High	Low
Integration	High	High	Medium
Testing and validation	Medium	High	Low
Operations	Same	Same	Same
Management	Medium	High	Medium

3.6.1.2 Hardware Costs

Hardware costs are also affected by the approach chosen. These costs can be broken into three categories: parts cost, waste and damage cost, and testing materials and supplies cost. Each will now be discussed.

Parts Costs

The level of parts costs that are incurred varies significantly between the three options. The framework approach and the build-from-scratch approach will have the lowest level of parts costs as both will involve buying small components and board fabrication services. The vendor kit approach will cost significantly more in this category as the parts procured (either as a single complete, excluding payload, kit or as individual subsystem components) will include amortized vendor development costs, vendor-incurred waste/risk costs, and vendor fabrication labor costs.

Waste and Damage Costs

Waste and damage costs are an inherent component of any development project. These costs will be particularly high for projects that must utilize parts for testing designs that may be deemed dysfunctional and thus require rework and the disposal of components that cannot be salvaged. Damage costs are increased by inexperienced developer staff who may inadvertently break parts during assembly or testing. It is expected that the develop-from-scratch approach will generate the highest level of waste and damage costs. A developer using the framework approach will enjoy some savings versus the develop-from-scratch approach because the designs have already been validated reducing the level of development and redevelopment required. The fabrication instructions should also have a positive impact as they should decrease the level of errors that occur during fabrication.

The vendor kit approach may decrease waste costs, as the need for consuming parts for testing is eliminated (except for the payload components). Damage costs should be reduced due to the fact that most assembly is conducted by the vendor (and thus their cost, which is factored in to their pricing). However, if damage occurs during integration or testing, it could be particularly expensive as an entire assembly may need to be replaced, if damaged.

Testing Materials and Supplies Costs

Materials and supplies may need to be consumed for testing purposes. The amount of cost incurred for these should be small, relative to the total cost of the spacecraft. It is likely that more testing materials and supplies will be needed for the framework and build-from-scratch approaches, as testing will be required at the component and assembled subsystem level in addition to the integration testing that will be required for all approaches.

3.6.1.3 Lab Equipment Costs

The impact of the approach chosen on lab equipment costs is more difficult to quantify as it depends on several factors. Chief among these is what lab equipment the developers already possess. A developer (such as a major university) that already has electrical and mechanical engineering laboratories may already possess most if not all of the laboratory equipment required for any approach. Further, any equipment that is procured may serve multiple uses and certainly could serve multiple spacecraft missions. Given this, the equipment may not be procured using mission funds or the mission may only be required to contribute a portion of the overall expense. Due to the wide variety of possible outcomes, this category (and launch costs, which are not directly affected by the approach selected) will be treated as neutral. The lab equipment costs may have an impact for a given developer (and should be considered in this context); however, their impact cannot be suitably generalized.

3.6.1.4 Summary

The type and cost of labor being utilized may be the driving decision factor in terms of what approach is undertaken. Those with high labor costs and who don't need to modify standard subsystems may find a vendor kit to be a preferred solution, as the labor costs required to fabricate a spacecraft from parts may exceed the vendor amortized development costs, assembly costs, and profit margin built in to the cost of a kit. Those with lower levels of labor costs (e.g., those using student workers, volunteers, or participants for academic credit) may find fabrication from parts to be a more prudent choice. If substantial changes are required or the base framework design is not suitable for the developer's needs, then the development-from-scratch approach may be most prudent. Table 3.2 presents a comparison of the level of each cost incurred for each category.

Table 3.2 Comparison of levels of category costs

	OPEN	From-scratch	Vendor kit
Labor costs	Medium	High	Low
Equipment costs	Low	Low	High
Lab equipment costs	Same	Same	Same

3.6 Qualitative Evaluation of the Value of the Approaches 49

3.6.2 Consideration of Recurring Amortized Vendor Development Costs

The vendor amortized cost has been discussed as part of the general topic of cost reductions, in the directly proceeding section. The prospective value of removing them on a recurring basis bears special consideration. Swartwout [28] proffers that a number institutions have had difficulty initiating a second small spacecraft program after the completion of their first. Institutions that develop a spacecraft from scratch face a significantly lower barrier to a second spacecraft, as compared to those who buy a vendor kit. The first spacecraft incurs the design costs and significant testing costs related to identifying and rectifying design issues. This may impair the development completion of this first craft, but it enables future ones. Of course, the later program must still incur hardware, assembly, and testing costs; however, the overall cost is significantly less than that incurred by the first spacecraft. Even if changes need to be made (due to a changed mission or other factors), the work on the first mission reduces the costs of the subsequent ones. Framework use allows an approximation of this, starting with even the first mission, as many development costs can be avoided and testing and other costs reduced. Of course, the development of local competency and a trained staff will not occur immediately; however, the costs can be significantly lower than with a vendor kit.

For commercial reasons, the level of profit enjoyed by vendors (over marginal unit fabrication and assembly costs) is not known. It would seem, however, that this must be a significant portion of the cost of the spacecraft (or components) given the parts costs previously presented [5] for the OPEN design and the legitimate need for the vendor to recover significant design and testing expenses over a small number of units.

The benefit of avoiding vendor costs will be enjoyed the most by those users with a greater number of missions. This will be particularly true for those with higher mission frequency levels, as repeated production will minimize retraining and other costs and increase fabrication speed (and thus lower costs) as staff gains experience.

3.6.3 Ease of Modification and Extensions of Design

Presuming that all of the designs and testing plans for the framework are freely available (as will be the case with OPEN), in addition to providing a turn-key solution, it provides an excellent starting point for developers who seek to create a CubeSat with unique capabilities. With a kit-based approach, the developers would likely, at a minimum, be required to redevelop (from scratch) the subsystem that they desired to alter and integrate this with other vendor-supplied spacecraft components. Alternately, they could pay the vendor to develop the new or modified component. With the framework, the developer can start from the known-good design documents, fabrication instructions, software and testing plans, and make the changes that are required to adapt the particular piece of hardware to the developer's requirements. The level of cost savings attained will, of course, depend on the magnitude of changes required. The savings can be conceptualized through identifying the percentage of the subsystem or component that is left unchanged. Obviously, the ability to customize pre-existing designs results in little value for a component or subsystem that is being completely redesigned; however, this value is more significant when only a small change or addition to a subsystem is required.

3.6.4 Allowing Focus on Area of Interest

The framework facilitates the construction of a partially modified CubeSat by developers that may seek to perform subsystem development or an engineering experiment which focuses on a single subsystem. Using the framework allows the researcher (or student project group, etc.) to lower project cost (as opposed to buying components) and also have the flexibility to use the framework subsystem as a starting point (or integration reference) for the custom-developed subsystem. It also facilitates a gradual transition to a custom-developed spacecraft: framework components can be used and modified, as desired, or replaced with custom-designed components.

3.6.5 Benefits Related to Export Control (EAR/ITAR)

Both the International Trafficking in Armaments Regulations (ITAR) and the Export Administration Regulations (EAR) recognize (among others) two types of exemptions: fundamental research and public domain. The former exempts items and documentation created as part of a university research project, subject to certain limitations. The latter exempts documentation of a type that would normally be available at a library or at a conference open to all (technically qualified) individuals. The former exemption is more helpful, as it has been taken to include both the technical documentation and the actual hardware. If compliant with the fundamental research exemption, framework documentation and hardware (produced at academic institutions, subject to the limitations stated in 22 CFR Chapter I, Subchapter M 120.11(a)(8) and 15 CFR §734.3 and §734.8) may be able to be made available to non-United States nationals [29]. This allows foreign students to be included in small spacecraft development projects and for the framework materials and the spacecraft components to be used for educational purposes in classes that contain foreign nationals. Proposed ITAR changes which change the public domain exemption may further increase the importance of the university research exemption.

3.7 Mix and Match

In some cases, it may be possible to combine parts (and benefits) of several of the approaches above. The limited use of a framework, for example, may facilitate rapid design and development of certain parts for an otherwise largely bespoke system. Alternately, parts from a kit/parts vendor may be used for this same purpose. If a developer is able to stay interoperable with the preexisting framework or kit designs, then the use of the purchased hardware or existing designs may be quite straightforward and effective. It may also be possible to adapt (using custom hardware) parts from one vendor (or framework developer) to be interoperable with another provider's hardware or designs. This would allow a bespoke system to draw on multiple sources of hardware and designs or a kit system to use components designed for a framework (or vice versa). The value of a mix-and-match approach, thus, must be clearly assessed within the context of the particular mission and costs and benefits of the hardware being considered for use.

3.8 Conclusions

This chapter has provided an overview of several different approaches to small spacecraft development, ranging from procuring a largely prebuilt kit spacecraft to which payload-specific components can be added to bespoke development. It has considered the value of each prospective approach and their possible combination. The key element of this decision making is, of course, the needs of the particular mission. Thus, systems such as those proposed by Wertz et al. [6] and Fortescue and Swinerd [7] should be utilized to define mission objectives, requirements constraints, and a mission design that can inform the build versus buy (and to what extent) decision.

References

1. Straub, J., C. Korvald, A. Nervold, A. Mohammad, N. Root, N. Long, and D. Torgerson. 2013. OpenOrbiter: A low-cost, educational prototype CubeSat mission architecture. *Machines* 1: 1–32.
2. Straub, J., and D. Whalen. 2014. Evaluation of the educational impact of participation time in a small spacecraft development program. *Education Sciences* 4(1): 141–154.
3. Straub, J. 2013. Increasing national space engineering productivity and educational opportunities via intrepreneurship, entrepreneurship and innovation. *Technology and Innovation* 15: 211–226.
4. ———. 2012. Cubesats: A low-cost, very high-return space technology. In *Proceedings of the 2012 Reinventing Space Conference*. Los Angeles, CA.
5. Berk, J., J. Straub, and D. Whalen. 2013. Open prototype for educational NanoSats: Fixing the other side of the small satellite cost equation. In *Proceedings of the 2013 IEEE Aerospace Conference*. Big Sky, MT.

6. Wertz, J.R., D.F. Everett, and J.J. Puschell. 2011. *Space Mission Engineering: The New SMAD*. Hawthorne, CA: Microcosm Press.
7. Fortescue, P., G. Swinerd, and J. Stark. 2011. *Spacecraft Systems Engineering*, 4th ed. West Sussex: Wiley.
8. Straub, J., and R. Fevig. 2012. Formalizing mission analysis and design techniques for high altitude ballooning. Presented at Proceedings of the 3rd Annual High Altitude Conference.
9. Straub, J. 2015. In search of technology readiness level (TRL) 10. *Aerospace Science and Technology* 46: 312–320.
10. *Definition of Technology Readiness Levels*. http://esto.nasa.gov/files/trl_definitions.pdf.
11. *Web Squared: Web 2.0 Five Years On*. http://gossgrove.com/sites/default/files/web2009_websquared-whitepaper.pdf.
12. *Sputnik and the Dawn of the Space Age*. http://history.nasa.gov/sputnik/expinfo.html.
13. *Explorer-1 and Jupiter-C*. http://history.nasa.gov/sputnik/expinfo.html.
14. May 2011. *Commercial Space Transportation Forecast*. http://www.faa.gov/about/office_org/headquarters_offices/ast/media/2011%20Forecast%20Report.pdf.
15. Shiroma, W.A., L.K. Martin, J.M. Akagi, J.T. Akagi, B.L. Wolfe, B.A. Fewell, and A.T. Ohta. 2011. CubeSats: A bright future for nanosatellites. *Central European Journal of Engineering* 1(1): 9–15.
16. Swartwout, M. 2004. University-class satellites: From marginal utility to 'disruptive' research platforms. Proceedings of the 18th Annual AIAA/USU Conference on Small Satellites. 11pp.
17. ———. 2009. The promise of innovation from university space systems: Are we meeting it? Proceedings of the AIAA/USU Conference on Small Satellites.
18. Woellert, K., P. Ehrenfreund, A.J. Ricco, and H. Hertzfeld. 2011. CubeSats: Cost-effective science and technology platforms for emerging and developing nations. *Advances in Space Research* 47(4): 663–684.
19. *Price List*. http://www.tyvak.com/products/Pico/TyvakPriceList_v1.05.pdf.
20. *CubeSat Personal Satellite Kit*. http://www.interorbital.com/CubeSat_1.htm.
21. *Pico-Class Products Price List*. http://www.tyvak.com/products/Pico/TyvakPriceList_v103_.pdf.
22. Straub, J., and R. Fevig. 2012. Achieving educational outcomes through CubeSat curriculum incorporation. Proceedings of the 9th Annual Cubesat Workshop.
23. *MSU Satellite Orbits the Earth After Early Morning Launch*. http://www.montana.edu/cpa/news/nwview.php?article=10458.
24. *Explorer 1 Overview*. http://www.nasa.gov/mission_pages/explorer/explorer-overview.html.
25. Twiggs, R., and B. Malphrus. 2011. CubeSats. In *Space Mission Engineering: The New SMAD*, ed. J.R. Wertz, D.F. Everett, and J.J. Puschell, 803–821. Hawthorne, CA: Microcosm Press.
26. Straub, J., J. Berk, A. Nervold, R. Marsh, and D. Whalen. 2013. The open prototype for educational NanoSats. *University of North Dakota Graduate School Scholarly Forum*.
27. Straub, J., J. Berk, A. Nervold, C. Korvald, and D. Torgerson. 2013. Application of collaborative autonomous control and the open prototype for educational NanoSats framework to enable orbital capabilities for developing nations. In *Proceedings of the 64th International Astronautical Congress*. Beijing, China.
28. Swartwout, M. 2011. AC 2011-1151: Significance of student-built spacecraft design programs it's impact on spacecraft engineering education over the last ten years. Presented at Proceedings of the American Society for Engineering Education Annual Conference. http://www.asee.org/file_server/papers/attachment/file/0001/1307/paper-final.pdf.
29. Straub, J., and J. Vacek. 2013. Escaping earth's orbit but not earthly regulations: A discussion of the implications of ITAR, EAR, FCC regulations and title VII on interplanetary CubeSats and CubeSat programs. In Proceedings of *the Interplanetary CubeSat Workshop*. Ithaca, NY.

Chapter 4
Starting a Small Spacecraft Program: Types of Programs and Their Benefits and Drawbacks

The previous two chapters have considered important foundational questions regarding the formation of a small spacecraft program. Chapter 2 considered why institutions might seek to start a small spacecraft program. Chapter 3 assessed several different approaches to the question of what type and scale of a program to form, in terms of a key question: whether to design and build a spacecraft from scratch, buy a vendor kit, or take a hybrid approach. This brief chapter deals with yet another foundational question for small spacecraft program formation: what the focus of the program will be. For an academic institution, this could be one of four principle types: a research program designed to (a) reach internal goals or (b) goals of a partner entity. The program could, alternately (c) focus on only educational pursuits. This, of course, was the original goal of the CubeSat form factor [1]. Alternately, it could (d) seek to combine both research and educational goals.

Commercial, civilian government or military, entities may not have the same focus on educational pursuits (though, some may see an educational value of a small spacecraft program for workforce development). They may seek to conduct research activities (either their own or, particularly in the case of a commercial firm, those of another party) or conduct non-research operations.

This chapter considers all four program types that would be typical for academia. An overview of each is presented in the sections that follow. First, relevant background material is presented in Sect. 4.1. Sections 4.2 and 4.3, which deal with research programs, are applicable to those within and beyond academia. Despite the terminology used, many of the considerations of these sections would be applicable to non-research operations conducted by commercial, governmental, military, or academic institutions. Sections 4.4 and 4.5 deal with educationally focused program types that would be more typical of academic institutions. Finally, Sect. 4.6 concludes the chapter with a discussion of the decision-making process applicable to an academic institution choosing between prospective program types.

This chapter is based on, revises and extends the paper "A Curriculum-Integrated Small Spacecraft Program for Interdisciplinary Education".

4.1 Background

Two areas of prior work are now reviewed. First, work on project-based learning and experiential education is reviewed. Then, an overview of small spacecraft development is presented.

4.1.1 Project-Based Learning and Experiential Education

Project-based learning (PBL), also commonly known as problem-based learning, experiential learning, or experiential education (EE), has been shown to be effective as a component of collegiate (and other levels of) education. The approach provides students with the opportunity to gain practical experience in a workplace-realistic setting, apply lessons learned through conventional lecture style education (thus gaining an appreciation for the importance of the material, internalizing it and increasing its retention), and to learn new skills which may fall outside of the boundaries of individual courses. These skills (e.g., project management, teamwork, cross-disciplinary collaboration techniques) may be important or more important to the students' long-term success than particular technical skills.

The implementation of PBL and EE in an educational environment can take several forms ranging from a completely PBL/EE-driven course (where objectives are attained via directed student tasks and inquiry) to the incorporation of limited duration PBL/EE exercises within the context of a more formally structured course. The former is typical of project-style courses, such as senior capstone projects, while the latter may serve to augment courses with specific skill-development focuses. Larsen et al. [2], for example, integrated small satellite development as a PBL component into a variety of undergraduate and graduate engineering courses. They state that the PBL content of these courses was approximately 50 %, with the remainder consisting of traditional style course activities. Hoic-Bozic et al. [3] show that a blended learning approach, consisting of PBL, collaborative and independent learning activities increased academic achievement levels and decreased the student dropout rate.

Okudan and Rzasa [4] proffer that PBL incorporation can be utilized to drive entrepreneurship in student participants. They review the results of an entrepreneurial leadership engineering course during which students had to develop and produce a product to sell. These results were largely positive and indicated that the course facilitated development in key areas including "leadership, motivation, innovation, communication skills, teamwork and writing business plans". It also was demonstrated to encourage entrepreneurial behavior in students. Doppelt [5] shows that PBL has benefits that go far beyond the classroom. This work showed that PBL with a "scientific-technological" focus increased student motivation and even their self-image. This work, in the context of middle and high school education was also shown to increase student performance on critical exams and even their college acceptance rates.

4.1.2 Small Spacecraft Development

The CubeSat form factor was developed by Robert Twiggs and Jordi Puig Suari in the late 1990s as a way to allow direct student participation in the development of a spacecraft [1]. By reducing the size of the spacecraft, project scope as well as development and launch costs were reduced. This allowed greater risk taking and more opportunity for student involvement and leadership. Numerous small spacecraft have been produced; Swartwout [6] has identified nearly one hundred universities who have successfully flown a mission. Many of these universities have conducted more than one mission and many more have missions under development.

The costs of small spacecraft development, however, are still outside the capabilities of many institutions. Complete development, from scratch, of a CubeSat may cost $250,000 or more [7]. Kits that provide all of the functionality required (except payload components) can cost as little as $40,000 [7]; however, this cost fails to consider payload development, integration, testing and other expenses. The use of a kit also may reduce student participation and limit innovation due to a need to conform to the integration requirements of the vendor producing the kit. Modification of a kit component, generally the proprietary property of the vendor, may require its redevelopment from scratch, making a minor change a significant expense. The replacement cost of these components may also have the effect of reducing acceptable risk levels and (desirable [8]) risk taking by students. Prior work [9] identified a variety of risk factors salient to this type of student project and presented a model for assessing the impact of these risks on project success. At a minimum, utilizing student workers (who may lack sufficient experience to assess the level of risk that they are taking or fail to realize that an action may damage a component) necessitates increased reserve funds for component replacement. This, again, raises project costs.

4.2 Internal Research Program

Perhaps the simplest type of small spacecraft program to describe, albeit the most versatile in format is that of an internal research program. The internal research program basically starts with a research question (or set of questions) and develops from there. Depending on the nature of the question, it may be possible to buy many of the spacecraft components from vendors or buy and modify a vendor kit. Typically, the mission-specific payload hardware components (or applicable subsystems, etc. in the case of a technology demonstration mission) will need to be developed from scratch. In other cases, cost pressures or experimental considerations will dictate that the entire spacecraft (as well as the payload components) be developed from scratch.

In this type of program, the focus will be primarily on attaining answers to the research questions; however, the goals may vary somewhat as the process allows them to be refined. This type of a program will typically have a research question (or technology demonstration) as a primary objective. Secondary and tertiary objectives may focus on other research questions related to the primary question or on (in the case of a primarily science mission) technology demonstration. While some educational benefit may be enjoyed by student (or other) participants, this is not an enumerated objective and not assessed.

4.2.1 Benefits

The principal benefits of the internal research program approach include flexibility, the ability to secure funding from multiple sources (including, for academic institutions, institutional funding as well as funding from regional, national and, in some regions, international organization sources) and compatibility with the organizational culture. Additional benefits include greater control and the ability for staff (in particular, lead investigators) to focus on areas of work that are of particular interest to them.

4.2.2 Drawbacks

Drawbacks to the internal research program approach include a prospective inability to secure funding (which is left to the lead investigators to secure, under this model) and the associated prospective lack of program longevity and ability to conduct long-range planning that having to secure support on a recurring basis entails. Additionally, this approach lacks the prospective technical assistance and topic guidance that would be available under partner-involved programs. It also makes program success entirely dependent on the nature and achievement of research goals (as opposed to diversifying between research, educational and other prospective objectives that may allow a partial success instead of a complete failure should research fail to answer a key question or the question/results be found to be uninteresting to the larger scientific community). Finally, the lack of identified and tracked educational benefits may limit the potential interest of students, increasing labor costs and, potentially, making finding workers difficult in an academic environment.

4.3 External Partner Research Program

The external partner research program is similar in form and format to the internal program, except that the research goal (at least the primary objective/goal) originates from outside the organization. For example, in an academic environment, the

4.3 External Partner Research Program 57

institution may be performing research for a commercial (or government) partner. In a corporate environment, the business may be performing services for another company or government institution.

The practical implications of this type of program fall into several categories. From a funding perspective, the external partner may be providing all (or most) of the funding. This may facilitate program longevity and prevent investigator time from having to be recurrently devoted to seeking funding to enable the program. Reliance on the external partner for funding, on the other hand, may make the program beholding to the political, logistical and financial circumstances of the partner organization (and of the program sponsors within the organization). Depending on the legal disposition of the intellectual property produced by the program, it may be difficult to transfer an externally-partnered program to an alternate funding model.

From a control perspective, the external partner may have significant influence over the direction and goals of the program and, in some cases, even a say in day-to-day decision making. This may exclude the ability to follow up on unexpected developments (even via an alternate funding mechanism, if the primary partner is unwilling to support the work, due to intellectual property ownership considerations).

From an environmental perspective, the needs of the external partner may drive schedule and other requirements that are atypical to the work-performing organization (particularly in the case of a typically relaxed academic research organization). This may create culture conflict between the two organizations.

For some organizations, policies may dictate certain boundaries between sponsor and performing organizations that may impact the ability to perform the work in certain ways. They may also create paperwork and necessitate time being devoted to other bureaucratic procedures. Policies may also preclude certain types of collaborative ventures and the potential for policies to be changed or re-interpreted may generate mission uncertainty, particularly for longer-term mission models.

4.3.1 Benefits

The principal benefit of the external partner research program is the external partner. Partners prospectively bring with them a number of benefits. First, the partner (who is presumed to be supplying the research question that they wish to fund the answering of) supplies an inherently relevant question, as the question likely will have direct relevance to the partner's business (or government/military entity operations). This will remove uncertainty regarding the possible reception of the completed work by the larger community (as there is a demonstrated need for it). Second, the partner may present a more reliable source for long-term funding for a mission (or multi-mission series), presuming that its research needs do not change significantly and appropriate progress is made in answering them. Third, the partner may evoke a special interest in student participants who may see the partner as a prospective employer that they paid or unpaid involvement in the project provides an opportunity to demonstrate their capabilities to. Finally, the partner may be able to provide technical, logistical and other assistance that facilitates allowing the researchers to focus solely on their key questions (instead of having to deal with the additional scaffolding that may be required to support the research, in the absence of the partner).

4.3.2 Drawbacks

The partner approach also presents several drawbacks. The first potentially problematic aspect of the partnership may be its creation. Both the performing institution and prospective partner may have policies which may lead to conflict on a multitude of topics ranging from the ownership of intellectual property to workplace conditions to facility access. The resolution of these conflicts, if possible, may lead to a variety of restrictions being placed on a project that may impair certain types or styles of work or otherwise interfere with a preferred working approach. Additionally, partner restrictions may change the treatment of the work under the International Trafficking in Arms Regulations (ITAR) and Export Administration Regulations (EAR) [10–12], leading to additional possible restrictions and logistical considerations.

The second prospective drawback to working with the partner may be workstyle differences. Academic institution investigators may be unaccustomed to the drive to produce results of the non-academic partner or the evaluation of the work based on a single success metric relating to the business-relevant research question. Like with the institutional program, the lack of identified educational goals and tracking may reduce student interest and involvement. Partner schedules may make the work incompatible with the academic semester/quarter/term-based scheduling approach. Partner funding conditions may also place restrictions on the use and/or dissemination of the results of the work.

4.4 Education-Only Program

The education-only program places its focus on student learning. This may take several forms. The most basic may be to simply have students produce a spacecraft for coursework reasons. This spacecraft (or a subset of a spacecraft) may never be designed to be launched; however, its construction will develop students' knowledge and skills regarding design, fabrication, testing and other areas. A second approach may be to have a student-generated science mission. This student-generated mission may support additional pedagogical goals related to mission analysis and design; however, mission funding and the determination of its success or failure would not be tied to the performance of the student-determined mission, but rather to the educational goals.

4.4 Education-Only Program

Education-only missions may lack many of the funding sources that other missions types may be able to access. At academic institutions, these types of activities may need to be supported out of student fees or teaching funds. Corporations, government/military institutions and various other organizations may support missions (or components of missions) for workforce development purposes—to train their staff or members in new techniques for spacecraft development. These types of missions could also be used in the corporate environment to assess the capability of prospective hires. United Launch Alliance (ULA), for example, has used small-scale rocket development to assess the capabilities of intern employees to assess their suitability for careers at ULA.[1]

These types of missions fill the need, in the context of collegiate education, to provide students with hands-on experiences that enrich their learning and aid in material understanding and retention. These experiences can be simplistic and limited in scope to an individual subject or lesson or they can be broad based and provide the students with learning opportunities outside of the scope of the particular subject at hand. A course in electrical engineering might, for example, teach students time management skills, in addition to circuit design and analysis skills, and a computer science course could provide students with leadership opportunities.

In both of these cases, however, students are still well within their comfort zone. They are working with their classmates (many, for those in the later years of a degree program, of whom they have taken multiple classes with). These are also, generally, students who speak the same field-of-study language. Hotaling et al. [13] suggest that this may not be the best approach, as students gain significant benefits from working in an interdisciplinary environment. In this type of environment, students gain skills that are valued by employers and which cause students to be more employable (as indicated by the percentage receiving a position upon graduation [13]).

Small spacecraft development programs allow students to gain experience in an interdisciplinary work setting that would be similar to the cross-disciplinary environments that many would be seeking employment in. Feedback from participants indicated that in addition to meeting interdisciplinary experience and technical learning goals, numerous additional educational benefits were enjoyed [14]. These included leadership skill development, providing perspective with regards to the importance of certain educational areas, improving writing skills, and generally increasing student confidence.

4.4.1 *Benefits*

The education-only program presents several prospective sources of benefit. The first is the use of educational objectives to justify program formation and demonstrate success. Prior work (which is discussed in later chapters) demonstrates the utility of this.

---

[1] See, e.g., http://www.ulalaunch.com/ula-and-ball-aerospace-2015-student-rocket.aspx

For the work in the OpenOrbiter program at the University of North Dakota, for example, results demonstrated improvement, on average, in all identified categories. The results also show that the improvement was practically significant (between 1 and 2.5 points on a nine-point scale—representing an 11–28% improvement) for those who showed improvement. This improvement was also attributed to program participation in all cases, with this attribution being strongest for technical skills and space interest.

The student responses to this survey (see Figs. 4.1 and 4.2) indicated that the program had met its goals within a short timeframe. Other results [14] indicated a strong correlation between serving in a leadership role and the level of benefit received and between the duration of participation and the level of benefit received. Also, anecdotal evidence suggests that a variety of other benefits, related to the central learning themes of the program, were enjoyed by participants.

These types of results are easy to quantify (using survey mechanisms and other approaches) and, for a well-formulated program, likely to occur. Unlike a research question (to which the attainment of an answer, much less the utility and practical significance of the answer) is inherently unknown, educational programs have demonstrated approaches that have been previously shown to attain results.

The education-only program may also allow access to institutional teaching funds that are not available to other program types. The demonstrated benefit to students may drive student interest in participation. Alignment with the teaching goals of the academic institution also reduces the likelihood and severity of culture conflict.

Perhaps most importantly is the fact that the program can be tailored to maximize the benefits to the student participants. This allows them to get the most educational benefit for the time that they put into the program and may have significant broader impact from the potential career possibilities that it unlocks for them (and the impact of their later work).

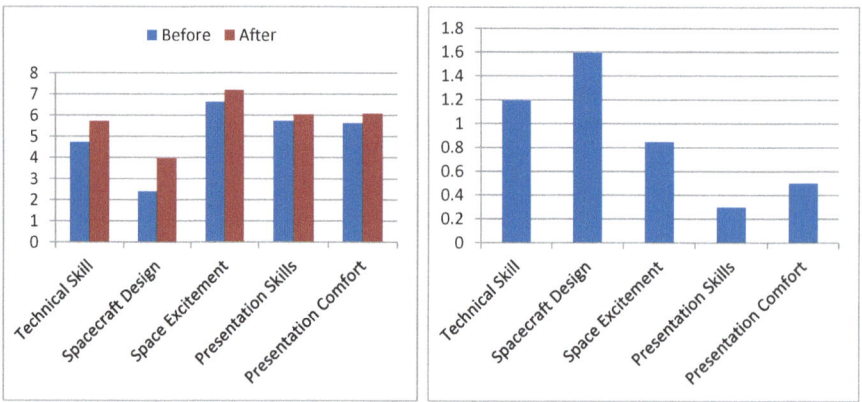

Fig. 4.1 (**a**) Comparison of reported pre-participation skill/comfort/excitement level and post-participation level, [14], *left*. (**b**) Average improvement in skill/comfort/excitement level, [14], *right*

4.5 Hybrid Research Education Program

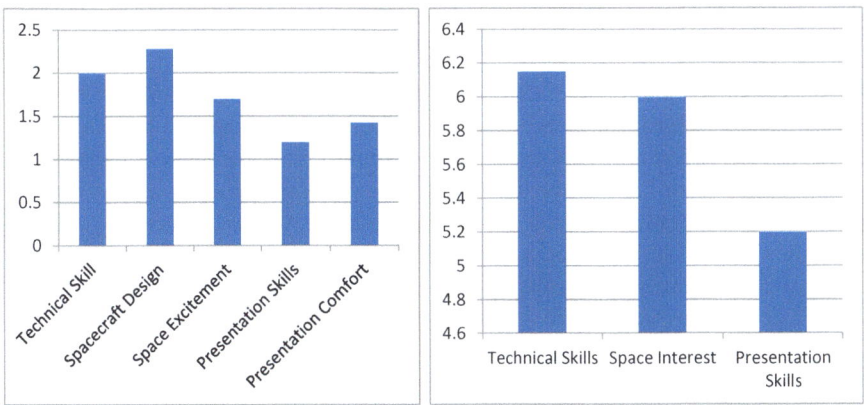

Fig. 4.2 (**a**) Average improvement in skill/comfort/excitement for students showing improvement in each category [14], *left*. (**b**) Attribution of benefit to program participation [14], *right*

4.4.2 Drawbacks

Unlike the research-focused projects, the education-only project may lack a key component to secure the interest and time of faculty and research staff. The lack of a publishable science or engineering objective may reduce the prospective value to those whose jobs require regular publications (particularly if publications related to student outcomes, if possible, are not valued by the institution). This is, of course, despite the fact that the educational program may take the same (or even more) time to run. The lack of a science/engineering objective may also impact student interest and the desire to be involved. It may also impair the value of the work to student participants who no longer have the research-based project to discuss with prospective employers (or the employer-connection provided by the partner approach). A student-generated research topic may mitigate aspects of this; however, the utility of the experience may then be (at least partially) driven by the quality of and scientific community interest in the topic selected.

4.5 Hybrid Research Education Program

Many programs will incorporate portions of each of the three aforementioned program types. They may, for example, have primary (and even secondary) research (or operational) benefits, but also incorporate secondary or tertiary educational benefits. Alternately, a program may have a partial external partner or have a primary educational objective but secondary or tertiary research ones. The hybrid model allows programs to attain benefit from all of the aforementioned areas, but may also subject them to some of the drawbacks of each, as well.

The benefits and drawbacks of each of the previously described approaches should be considered in formulating a hybrid approach program. Additionally, synergies and combination problems from combining aspects of each approach should be considered.

4.6 Academic Institution Decision-Making Process

In an ideal situation all prospective types of programs would be possible at a given institution and the pros and cons of each could be assessed using the previously enumerated benefits and drawbacks as a starting point for consideration. Pragmatically, most programs will fall into the hybrid category and thus require a combination of the benefits and drawbacks from several categories as well as the identification of numerous benefits and drawbacks specific to the various combinations that would be possible. In addition to the material presented in this chapter, the material in Chaps. 2 and 3 related, respectively, to why a program might be formed and the build/buy decision may be relevant to decision-making consideration. Additionally, the student-related risk factors discussed in Chap. 7 may be of significant interest in determining whether to combine educational and research (or other noneducational) goals. Timeframe considerations as well as funding and other logistical considerations may also be relevant, perhaps even to the point of dominating the decision. Some approaches may offer the ability to make changes to the selected model at later points and the value of this flexibility (particularly in an uncertain funding, legal, and logistical environment) should not be discounted.

4.7 Summary

This chapter has provided a guide to a key step in the process of starting a small spacecraft program: namely determining what type of program to start. To this end, the various possible program types have been discussed and benefits and drawbacks for each have been considered. It has been suggested that hybrid programs may offer both synergistic benefits (as compared to other single-category approaches) and pose problems caused by their combination (that are not faced by any single-category approach). Practically, as logistical considerations will likely drive many programs to have a hybrid approach, this chapter provides a list of prospective considerations for program formation to inform program founder decision making.

References

1. Deepak, R.A., and R.J. Twiggs. 2012. Thinking out of the box: Space science beyond the CubeSat. *Journal of Small Satellites* 1(1): 3–7.
2. Larsen, J.A., J.F.D. Nielsen, and C. Zhou. 2013. Motivating students to develop satellites in problem and project-based learning (PBL) environment. *International Journal of Engineering Pedagogy* 3(3): 11–17.
3. Hoic-Bozic, N., V. Mornar, and I. Boticki. 2009. A blended learning approach to course design and implementation. *IEEE Transactions on Education* 52(1): 19–30.
4. Okudan, G.E., and S.E. Rzasa. 2006. A project-based approach to entrepreneurial leadership education. *Technovation* 26(2): 195–210.
5. Doppelt, Y. 2003. Implementation and assessment of project-based learning in a flexible environment. *International Journal of Technology and Design Education* 13(3): 255–272.

References

6. Swartwout, M. 2013. The long-threatened flood of university-class spacecraft (and CubeSats) has come: Analyzing the numbers. Presented at Proceedings of the 27th Annual AIAA/USU Conference on Small Satellites.
7. Straub, J. 2012. Cubesats: A low-cost, very high-return space technology. In *Proceedings of the 2012 Reinventing Space Conference*. Los Angeles, CA.
8. Swartwout, M. 2004. University-class satellites: From marginal utility to 'disruptive' research platforms. Presented at Proceedings of the 18th Annual AIAA/USU Conference on Small Satellites.
9. Straub, J., R. Fevig, J. Casler, and O. Yadav. 2013. Risk analysis & management in student-centered spacecraft development projects. In *Proceedings of the 2013 Reliability and Maintainability Symposium*. Orlando, FL.
10. Straub, J., and J. Vacek. 2013. Do we have an ITAR problem: A review of the implications of ITAR and title VII on small satellite programs. Presented at Spring 2013 CubeSat Workshop.
11. ———. 2015. in press. Reforming regulation of basic and small business research and education in space technologies under the international traffic in arms regulations and the export administration regulations. *Journal of Space Law* 39:4.
12. ———. 2013. Escaping earth's orbit but not earthly regulations: A discussion of the implications of ITAR, EAR, FCC regulations and title VII on interplanetary CubeSats and CubeSat programs. In *The Interplanetary CubeSat Workshop*. Ithica, NY.
13. Hotaling, N., B.B. Fasse, L.F. Bost, C.D. Hermann, and C.R. Forest. 2012. A quantitative analysis of the effects of a multidisciplinary engineering capstone design course. *Journal of Engineering Education* 101(4): 630–656.
14. Straub, J., and D. Whalen. 2013. An assessment of educational benefits from the OpenOrbiter space program. *Education Sciences* 3(3): 259–278.

Chapter 5
Forming a Program: Funding and Organizational Issues

This chapter deals with some of the logistical aspects of starting a small spacecraft development program. It begins with a discussion of human resource needs, with particular attention being paid to student workers, as they are typically a key component of a university small spacecraft development program. Then, financial and other resource needs are discussed. Next, focus turns to strategies for organizing a small spacecraft program, before concluding.

5.1 Defining Resource Needs: Human Resources

A wide range of the so-called human resources will be needed for virtually any small spacecraft development program. The exact composition required will depend on a few factors. The first is the approach to spacecraft development that is selected. Four approaches were discussed in Chap. 3: bespoke, framework based, vendor kit, and hybrid. Typically, the bespoke approach will require experts with design, testing, and development skills in all areas of the spacecraft. The framework approach will reduce the design skill requirements in areas where no (or limited) modification to the framework is planned. It also reduces overall human resource needs due to supplying key design development and testing knowledge and walk-throughs. Skilled staff are, of course, still required for any areas where significant modification will be performed as well as, possibly, the areas affected by these areas. The vendor kit approach minimizes staffing needs. These individuals will principally focus on payload design and spacecraft integration and testing—areas requiring skilled staff under all approaches. The sections that follow discuss student human resource considerations. Specific areas of work are discussed in Sect. 5.3.

This chapter is based on, revises and extends the papers "Extending the Student Qualitative Undertaking Involvement Risk Model" [1] *and "OpenOrbiter: An Interdisciplinary, Student Run Space Program"* [2].

5.1.1 Student Involvement

Student involvement is a standard feature of university small spacecraft programs. While critical, in many cases to their success, student involvement carries with it certain challenges. If student involvement's desired benefit was solely student education, the need to characterize and mitigate risks would be dramatically reduced. A student or inexperience-specific or general risk factor's occurrence, however, can have impact to the student participant's success; it can also have a pronounced effect on the project as well. While students may gain (possibly even enhanced) benefit from risk actualization, the project stands to suffer. To characterize the magnitude of this impact, it is important to consider faculty perceptions of student involvement on research projects. Zydney et al. [3] proffer that faculty see students' participation as valuable, with over half of them indicating that students' contribution to their work was "important" or "very important." Thus, the failure of a student to make progress is a risk that may be comparable to causing damage or other types of impact on prior work.

While student participation is valuable to faculty, it appears that project completion may be less important to students, as [4] demonstrated a lack of correlation between the research productivity level of faculty and students' educational benefits.

5.1.2 Student Risk Perception

One reason that student workers may be more risk occurrence prone is a failure to properly assess risk likelihood and impact. However, despite a significant correlation between youth and inexperience, it is important to note the potentially confounding impact of risk perception. Because of this, there may be a performance difference between younger and older individuals with similar experience levels (i.e., typical age and older students) in a field. A full exploration of the topic of risk perception is far beyond the scope of this chapter; however, prior work on this topic is informative. Botterill and Mazur [5], for example, provide a general overview of the topic, while Slovic et al. [6] consider the value of studying it. Boholm [7] reviews and compares risk perception research over a 20-year period and Mitchell [8] considers risk perception and risk reduction in the context of an organization. The crux of the risk perception problem is that younger individuals may fail to appreciate the applicability of risk to them and its impact [9]. This has been documented across multiple areas, including driving [10], sexual [11], and other "health threatening" [12] behaviors. Steinberg [13] attributes the greater risk-taking tolerance of youth to "age differences in psychological factors that influence self-regulation." Thus, age may confound the experience/risk correlation and intensify certain risk factors when both young age and inexperience are applicable. Given this, traditional-age undergraduates may have a particularly higher propensity to fail to see how their actions, behaviors, or inaction may create risks, or the impact that these risks may have on them or others.

Risk perception, however, is not only affected by age. Correlation has been shown with gender [14], culture [15], and other factors [16, 17]. The impact of education in correcting risk perceptions has been demonstrated by Ronan and Johnston [18]. Weber and Milliman's [19] work suggests that "risk preference" may be a stable aspect of an individual's personality, highlighting the importance of risk perception on the acceptance or rejection of the risk in a given circumstance. Renn [20] discusses the importance of risk perception in relation to the management of risks.

Small spacecraft are in particular need of robust risk management mechanisms as they are commonly integrated as secondary payloads on rockets carrying other orders-of-magnitude more expensive hardware. They must meet the same (or perhaps even more stringent) integration standards as the primary payload. Some small spacecraft have also been launched via the International Space Station, necessitating their compliance with human safety standards. Once they are in orbit, they are also on their own, with no practical servicing capability. Design and implementation failures can, thus, cause a spacecraft to fail integration testing and not get launched, to fail subsequent to integration and damage expensive equipment or pose a threat to astronauts or fail on orbit, impairing mission performance. The training and research provided by these efforts is integral to developing new technologies as well as training the next generation of aerospace professionals. Given this, a better understanding of the risks posed by student and inexperienced staff involvement is necessary. In prior work [21], a risk model specifically targeted at students (and to some extent, at all inexperienced workers) was presented, called the Student Qualitative Undertaking Involvement Risk Model (SQUIRM). Chapter 7 provides a more in-depth look at student-involved project risk and its prospective mitigation.

5.2 Defining Resource Needs: Financial and Other Resources

The exact nature of the mission will obviously drive the specific financial and other resource needs that are required. This section, thus, doesn't focus on quantifying these needs (which will vary, tremendously, from mission to mission); instead, it presents a model to aid in the identification of applicable costs and their attribution to the correct category. This model is presented in Fig. 5.1 and will now be discussed.

The model identifies four key cost areas that combine to create the overall mission cost. These areas are spacecraft costs, general overhead costs, management reserve and contingency and mission operations costs. Several of these areas are multiple distinct component cost areas. Mission operations costs, for example, comprise staffing costs, general ground station facilities and equipment costs, and the costs of accessing or purchasing the requisite communications equipment for communicating with the spacecraft. General overhead costs comprise mission planning labor costs, management labor costs, general mission facilities and equipment (i.e., office space, etc.), intellectual property licensing costs (if any), and miscellaneous costs (a catchall category for other small overhead-type costs).

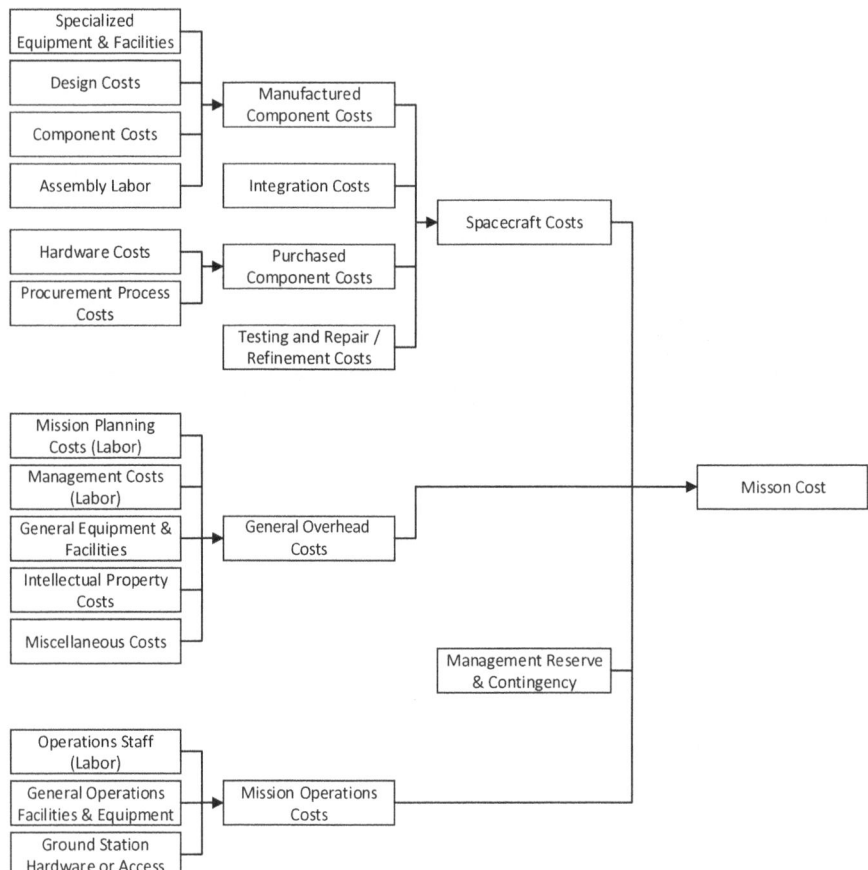

Fig. 5.1 Small spacecraft mission cost model

The spacecraft costs can be further subdivided into four subcategories: manufactured component costs, integration costs, purchased component costs, and testing and repair/refinement costs. Of these, two are further subdivided. The manufactured component costs include any specialized equipment or tooling needed, design costs, component (hardware) costs, and assembly labor. The purchased component costs include hardware costs as well as process costs (such as labor).

5.3 Organizational Strategies for Program Formation

The OpenOrbiter program will be used as a model for program formation. Many will want to reduce the complexity or add particular roles to meet their own needs, of course. OpenOrbiter is structured around a multilevel organizational hierarchy. The program is led by a program director and deputy program director. Initially, reporting to these individuals were four associate directors (for electrical, software, architecture, and communications, outreach, and policy) and three managers (for mechanical, operations, and ground station). Three of the associate directors had managers reporting to them; these managers include ground station software, operating software, payload software, sensors and bus, optical systems, power, electrical communications, group communications, outreach, and policy. Each associate director and manager is advised by a faculty mentor. These faculty mentors are from various departments spanning multiple colleges at the University of North Dakota.

At the onset, program management occurred via weekly meetings between each manager and his or her group members, and a weekly meeting of all team leads. Associate directors and managers were asked to send out a weekly e-mail to their team members and the communications team summarizing current tasks in progress. The communications team was tasked with creating a summary version that is sent to all participants. Associate directors and managers also have frequent contact with their faculty mentors, on an as needed basis. Approximately 300 students and 20 faculty members were involved directly or indirectly with the program. This organizational structure is shown in Fig. 5.2.

These initial student participants were solicited through in class presentations across a variety of disciplines and short informal seminars. These presentations were largely given by the associate directors to find managers and/or fill their respective teams. A number of orientation sessions were held to introduce prospective participants to the various opportunities for participation.

Over time, the initial large group became a core group which was doing most of the work on the spacecraft. Many of the initial procedures, needed for managing a large organization, were relaxed. Meetings, in particular, became smaller, ad hoc, and task focused.

5.3.1 Program Implementation

This section focuses on the components of a program. It uses the initial organization structure of the OpenOrbiter program as a guide to future program development.

5.3.1.1 Mission Design and Architecture

The mission design and architecture team focuses on high level decision making for the mission and the spacecraft. The team began by developing a high level mission architecture. This architecture document was circulated to members of all of the other teams to inform their design activities. Once the document was done, the mission design and architecture team served as a coordinating group between the other teams. For example, it coordinated changes made by other groups and ensured that mission requirements were adhered to and constraints were not violated. This group also resolves conflicting changes and is responsible for disseminating an updated mission architecture document to the other groups.

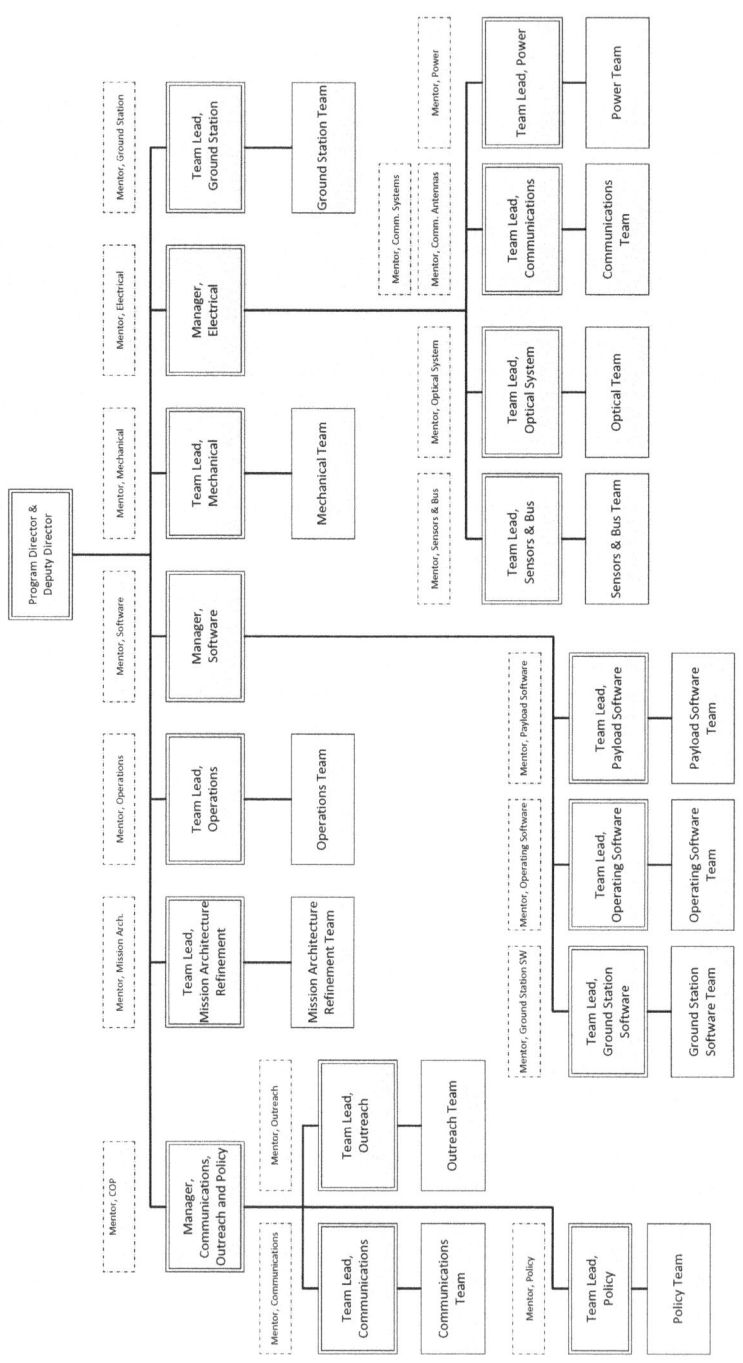

Fig. 5.2 OpenOrbiter organizational structure [2]

5.3.1.2 Communications, Outreach, and Policy

The communications, outreach, and policy group served a plethora of interrelated purposes. It was responsible for both internal and external communications (messages to team members, messages to media, etc.), outreach activities (e.g., community involvement), and policy considerations. Policy, in this context, includes considering the program's implementation within the university environment (how it is effected by and affects other university endeavors), funding needs, and integrating investigator areas of interest into a cohesive mission.

5.3.1.3 Electrical

The electrical team had four subgroups: sensors and bus, optical system, communications, and power. The sensors and bus subgroup focused on designing, developing, and testing the core system bus for the OPEN framework and the OpenOrbiter satellite. The subgroup also managed the integration and interaction between other electrical subsystems. The optical system group was responsible for the design, implementation, and testing of OpenOrbiter's optical payload. This group coordinated heavily with the payload software group. The communications group had one of the most challenging tasks in the project: developing a space quality radio. Finally, the power group was responsible for developing the power subsystems (including generation and regulation), as well as for the development of an architecture for maximizing power performance for the OPEN platform.

5.3.1.4 Mechanical

The mechanical team began by developing a novel structure (informed by previous work presented by Samson [22]) that allows the mounting of computer boards on the four interior faces, leaving a central area available for payload use. The mechanical team was also responsible for spacecraft mechanisms, thermal design, and associated testing.

5.3.1.5 Software

Three subteams existed within the software group: payload software, operating software, and ground station software. The payload software group was responsible for designing, developing, and testing mosaicking and super-resolution software that will comprise the primary payload of the OpenOrbiter spacecraft. A key mission goal of small satellite work at the University of North Dakota, at present, is to advance software and algorithms for mosaicking and super-resolution via incorporating IMU and GPS data to provide an initial rough alignment of images, reducing the computational requirements for these image processing technologies.

The next team, the operating software group, focused on designing, developing, and testing the software that would be used to control the spacecraft during on-orbit operations. This autonomous software was designed to respond to ground controller goals and commands, sensor input, system health status, and mission requirements to command spacecraft activities.

Finally, the ground station software team developed software to control radio-communications equipment to obtain data from the satellite, transmit data to the satellite, and perform processing of the telemetry received. This group worked in conjunction with the mission operations group to ensure that the software interfaces meet with that group's needs.

5.3.1.6 Operations

The operations team was responsible for the mission concept of operations. This group also had responsibility for overseeing the operations of the spacecraft on-orbit. The group integrated changes initiated by the architecture group or other groups into a revised mission concept of operations which is disseminated to all other groups.

5.3.1.7 Ground Station

The ground station team was responsible for the physical implementation of the ground station hardware, integration of the software with the hardware, and ground station operations (in conjunction with the operations team). This team investigated, selected, and tested possible ground station hardware solutions.

5.3.2 Implementation Difficulties

A brief discussion of implementation difficulties that were encountered is now provided. These examples and the solutions found are meant to be illustrative and not exhaustive. This section is provided in the hope that it may provide prospective solutions for others as well as a demonstration of a problem solving methodology.

5.3.2.1 Faculty Support

With a diverse group of skills required, identifying suitable faculty members to mentor project groups was a challenge. Once the initial prospective faculty mentors were identified, group leads approached them to act as a mentor for their group. Virtually all faculty members approached agreed to mentor the group. However, the level of time commitment possible from each faculty member varied greatly; some had to be replaced with more available faculty members. In one instance, two faculty members were selected to mentor a group, as a single faculty member with the requisite skill set could not be identified.

5.3.2.2 Divergent Interests of Faculty Participants

The faculty mentors, as would be expected, brought to the project their own research interests and involvement expectations. Team leads were trained to set clear expectations with faculty mentors regarding their role (advisory, not managerial) and the nature of the project. In some cases, synergies between project goals and faculty member research interests were identified, benefiting both. In other cases, regrettably, the project was unable to accommodate requests that were too divergent from its stated objectives.

5.3.2.3 Technical Challenges

A plethora of technical challenges were encountered, as would be expected. Virtually all of the students participating in the project began with little or no space domain experience (student expectations and starting and ending skill levels are discussed in Chaps. 8–10). Significant education regarding the basics of space mission design, mission operations, and other technical skills was provided. A key goal early on in the project was identifying and providing team leads with appropriate technical, management, and other documentation. A key consideration, raised by one of the faculty mentors, also early on in the project, was to identify a compatible scientific goal, in addition to the engineering one. While the key objective of the mission was to validate the OPEN framework, it would be difficult to show that this had occurred without demonstrating that the framework was able to meet the scientific objectives of a prototypical mission.

Remote sensing, thus, was a logical choice, given the faculty and research interests and skill sets available on the project team and at the university. A related project had previously focused on image processing; further developing this work was identified as a goal of the OpenOrbiter project. Several application areas within the capabilities of the projected optical sensor, which would provide an effective demonstration and test for the image processing software, were identified and selected as targets.

5.3.2.4 Logistical Challenges

Numerous logistical challenges were faced during the initiation and implementation of the OpenOrbiter program. The coordination of the numerous student and faculty participants was a significant undertaking. This was complexified by the physical separation between the aerospace college and the rest of the academic departments at the University of North Dakota. While problematic for management, dealing with the separation provided an analog for the separation typical in industry projects of this type.

Coordination was maintained through weekly team lead meetings, internally and public-facing Web sites, an internal communications program, frequent intragroup communications, and biweekly general meetings open to all participants.

5.4 Conclusion

This chapter has provided an overview of some of the logistical and other considerations and challenges encountered when initiating a small satellite program. While this chapter has relied largely on anecdotal evidence, it is hoped that it provides a notion of what to expect to those considering program initiation as well as an overview of areas to which to direct attention.

References

1. Straub, J. 2014. Extending the student qualitative undertaking involvement risk model. *Journal of Aerospace Technology and Management* 6(3): 333–352.
2. Straub, J., J. Berk, A. Nervold, and D. Whalen. 2013. OpenOrbiter: An interdisciplinary, student run space program. *Advances in Education* 2: 4–10.
3. Zydney, A.L., J.S. Bennett, A. Shahid, and K. Bauer. 2002. Faculty perspectives regarding the undergraduate research experience in science and engineering. *Journal of Engineering Education* 91(3): 291–297.
4. Prince, M.J., R.M. Felder, and R. Brent. 2007. Does faculty research improve undergraduate teaching? An analysis of existing and potential synergies. *Journal of Engineering Education* 96(4): 283–294.
5. Botterill, L., and N. Mazur. 2004. Risk and risk perception: A literature review. *Project no. BRR-8A, Rural Industries Research and Development Corporation, Barton.*
6. Slovic, P., B. Fischhoff, and S. Lichtenstein. 1982. Why study risk perception? *Risk Analysis* 2(2): 83–93.
7. Boholm, A. 1998. Comparative studies of risk perception: A review of twenty years of research. *Journal of Risk Research* 1(2): 135–163.
8. Mitchell, V. 1995. Organizational risk perception and reduction: A literature review. *British Journal of Management* 6(2): 115–133.
9. Weinstein, N.D. 1984. Why it won't happen to me: Perceptions of risk factors and susceptibility. *Health Psychology* 3(5): 431.
10. Deery, H.A. 2000. Hazard and risk perception among young novice drivers. *Journal of Safety Research* 30(4): 225–236.
11. Levinson, R.A., J. Jaccard, and L. Beamer. 1995. Older adolescents' engagement in casual sex: Impact of risk perception and psychosocial motivations. *Journal of Youth and Adolescence* 24(3): 349–364.
12. Cohn, L.D., S. Macfarlane, C. Yanez, and W.K. Imai. 1995. Risk-perception: Differences between adolescents and adults. *Health Psychology* 14(3): 217.
13. Steinberg, L. 2004. Risk taking in adolescence: What changes, and why? *Annals of the New York Academy of Sciences* 1021(1): 51–58.
14. DeJoy, D.M. 1992. An examination of gender differences in traffic accident risk perception. *Accident Analysis & Prevention* 24(3): 237–246.
15. Rippl, S. 2002. Cultural theory and risk perception: A proposal for a better measurement. *Journal of Risk Research* 5(2): 147–165.
16. Sjöberg, L. 2000. Factors in risk perception. *Risk Analysis* 20(1): 1–12.
17. Wildavsky, A., and K. Dake. 1990. Theories of risk perception: Who fears what and why? *Daedalus* 119(4): 41–60.
18. Ronan, K.R., and D.M. Johnston. 2001. Correlates of hazard education programs for youth. *Risk Analysis* 21(6): 1055–1064.

References

19. Weber, E.U., and R.A. Milliman. 1997. Perceived risk attitudes: Relating risk perception to risky choice. *Management Science* 43(2): 123–144.
20. Renn, O. 1998. The role of risk perception for risk management. *Reliability Engineering and System Safety* 59(1): 49–62.
21. Straub, J., R. Fevig, J. Casler, and O. Yadav. 2013. Risk analysis & management in student-centered spacecraft development projects. In *Proceedings of the 2013 Reliability and Maintainability Symposium*. Orlando, FL.
22. Samson, J. 2012. Update on dependable multiprocessor CubeSat technology development. Presented at 2012 IEEE Aerospace Conference.

Chapter 6
Forming a Program: Technical and Logistical Issues

The previous chapter discussed the management and funding of small spacecraft development programs. This chapter, now, considers how (presuming that these higher level challenges are met) one actually goes about starting a small spacecraft program. First, background details on reasons for forming a program are discussed. Next, focus turns to mission planning and goal selection. Following this, the key consideration of matching the program's goals and available funding is discussed. Then, technical and logistical decision-making strategies are discussed, with a focus on goal-based and requirement/constraint-aware decision making. Following this, the prospective technical models that could be used for small spacecraft programs are considered with a focus on design paradigm selection and a discussion of the appropriate level of design approach rigidity. Finally, a discussion of planning for program longevity is presented, before concluding.

6.1 Considering the Reasons for Forming a Program

A small spacecraft program could be formed to serve one of any number of technical, science, or engineering goals. Previous small spacecraft have been used or proposed for demonstrating enhanced imaging capabilities [3], new mechanical joint technology [4], for atmospheric studies [5] and a variety of other purposes. One key reason for small satellite development, for most university missions, is student learning. The CubeSat form factor traces its origin back to education, with Twiggs aiming to create a form factor that would let students finish a satellite within a constrained amount of time [6]. Swartwout suggests that student learning is, in fact, a key portion of being a university-class mission [7].

This chapter is based on, revises and extends the papers "Formalizing Mission Analysis and Design Techniques for High Altitude Ballooning" [1] *and "Evaluation of High-Altitude Balloons as a Learning Technology"* [2].

In the context of education, small satellites fall into a category commonly referred to as project-based learning (PBL) or experiential learning (EL). PBL and EL techniques have been shown to be effective in enhancing both student learning and excitement. Small satellites, as an educational technology, are well suited to PBL/EL learning. PBL and EE have been shown to be effective in engineering education [8, 9], aerospace applications [10, 11], and space mission design [12–14]. PBL/EE projects have been demonstrated at multiple grade and student-age levels [13, 15, 16]. Prior work has also demonstrated correlation between the amount of time involved in PBL/EE activities and the level of benefit enjoyed by students, in some cases [17].

PBL and EE have also been shown to produce a number of benefits in addition to aiding the learning of the targeted material [18]. These include driving-enhanced student creativity [19], motivation [20], and self-image [20]. They have also been shown to have a positive effect on student retention in an academic program [21], knowledge retention [22], preparation for joining the workforce [18], and job placement [23].

Mills and Treagust [24] sum up the challenge facing educators: they suggest that the "chalk and talk" approach to education isn't effective; however, PBL is difficult to evaluate, time consuming, and not well suited to traditional faculty review and promotion processes. Perhaps the expense of PBL/EL should be added to this list. Small satellites may be especially effective at driving student excitement; however, its cost levels exceed those of many other PBL/EE projects (including some with similar thematic focuses such as high altitude ballooning [2]). This chapter, in part, discusses how to maximize the value of the PBL/EE experience, for student participants. It also, importantly, provides information that will hopefully make a nascent small satellite project more successful.

6.2 Identifying Science, Technology Development, Educational, and Other Goals

The discussion in the previous section describes multiple reasons why an individual, group, educational institution, firm, or other entity may wish to start a small satellite program. The reason (or reasons) for program formation will drive virtually all other decisions. It is critical that the reason be well thought out and, possibly, refined to maximize the value of the program to the organization and program participants. These topics are discussed in this section.

6.2.1 Defining Objectives

The objective definition process can take a large variety of forms. In some cases, objectives may be highly influenced by a funding source or program mission statement (or program objectives). In other cases, objectives may have to be defined in an effort to seek funding (or other aid such as launch site access) and thus incorporate elements appropriate to this goal. In still other cases, requirements may be less constrained by funding and resource considerations.

The objective definition process should begin with stakeholder identification and a needs analysis. Stakeholder identification involves determining who is affected by a proposed activity. This includes individuals or entities that may fund the activity, those involved in the activity, and those that may be positively or negatively impacted by the activity, without direct participation. Once each stakeholder or group of stakeholders is identified (stakeholders with very similar needs should be grouped—if differences are found, then these groups can be subdivided, etc.), members of the group should be interviewed to determine their interest in the mission. Once a set of representative interviews has been completed, needs analysis should be conducted. Needs statements must then be refined into broad statements of objectives that are qualitative and easily understood. Again, the goal of objectives is to provide a general set of mission goals—not a quantitative set of requirements. The generated objectives should be shared with the stakeholders to ensure that they are in line with stakeholder's expectations and are understandable.

6.2.2 Maximizing Value

The effective use of small satellite development as a learning technology is based on five key principles. These will, generally, maximize the benefit that is achieved from satellite development incorporation. These five principles are common sense conclusions based on observations of successful and unsuccessful small satellite and ballooning learning activities.

1. Start with educational objectives (e.g., student learning about an engineering principle, learning about system integration, learning about command techniques) and add small satellite development, if it is relevant and aids this objective. There is little point to including an educational technology (including satellite development) that does not meet a salient educational objective.
2. Determine if satellite development is the best solution: satellite development fills a narrow niche. In some cases, the lower-cost high altitude ballooning may be a suitable alternative. Ballooning may not be suitable if needs include greater control over movement or longer duration flights. In some cases, UAVs may be better suited to these needs. If a higher altitude level is required or longer duration than can be supported by a UAV, then a satellite may be an excellent choice.
3. Once the technology use decision is made, maximize its value. While a UAV or high altitude balloon mission may take 6–8 h to conduct, plus numerous hours in preparation, a small satellite may require months of development and operations. Given the high level of commitment, all possible value should be attained. Additional value can be attained in two ways: increasing the value to each participant (e.g., adding in additional learning-goal-driven instructional modules that can be enabled by the launch) and expanding participation (e.g., inviting the participation from faculty and students from other disciplines to participate in a discipline-relevant way).

4. Design a small spacecraft mission; don't design a balloon, UAV, or large spacecraft mission and try to shoehorn it into a small satellite. The needs are different.
5. Maximize the small spacecraft-specific value: the particular characteristics of small spacecraft operations should be considered. For example, consider what elements of the space environment (that would not be present for a UAV or balloon mission) can be helpful. Also consider how an experiment could be spread over multiple small craft (see, e.g., [25–28]), instead of using a larger one.

6.2.3 Value Assessment

The assessment of the value of the educational benefit produced by incorporating small satellite development into a preexisting or new curriculum is based on the difference of the educational value produced by the small satellite development approach versus the conventional approach (and/or other alternatives under consideration). Assessment of the value of the conventional approach will vary depending on the particulars of the activity; however, a set of general guidelines is now considered.

The assessment process begins with the educational goals that the proposed activity should facilitate. Based on this, for the conventional approach (or other approach under consideration), the risks of the approach, the prospective learning benefits, and the enthusiasm benefits are considered (see Fig. 6.1).

Risks for classroom-based activities are generally, but not always (e.g., a laboratory activity with dangerous equipment), relatively low. They basically fall into three categories. First, the activity may fail to be completed. In this case no or limited educational benefit may be attained from it. Second, the activity may succeed, but fail to convey the desired educational benefits. In this case, despite the activity being completed as planned, students may receive no or limited educational benefits (as compared to the level desired). Third, the activity may have impacts beyond the level of learning that occurs. These may include injury to participants or others and/or damage to facilities or equipment. In this case, learning benefit may be limited (or eliminated) and residual problems created for the instructor or others. Risks can be assessed with a multistep process. First, they are identified. Second, their likelihood of eventuating is determined. Third, the impact of their occurrence is assessed. Forth, steps to mitigate their occurrence or impact level are developed. Fifth, a combined metric of the risk of occurrence and impact (in light of incorporated mitigation techniques) is produced.

The prospective learning benefits can be characterized in terms of factual, skill, and experience learning. Factual learning helps the students to learn, understand, or retain particular pieces of knowledge. Skill learning teaches students a process that they can repeat and, possibly, adapt to apply in other circumstances. Experience learning combines both facts and skills into a meaningful event where the knowledge and skills complement, reinforce, and assist with the retention of each other. The value of the learning, in the abstract, is difficult to assess (e.g., what might the

6.2 Identifying Science, Technology Development, Educational, and Other Goals

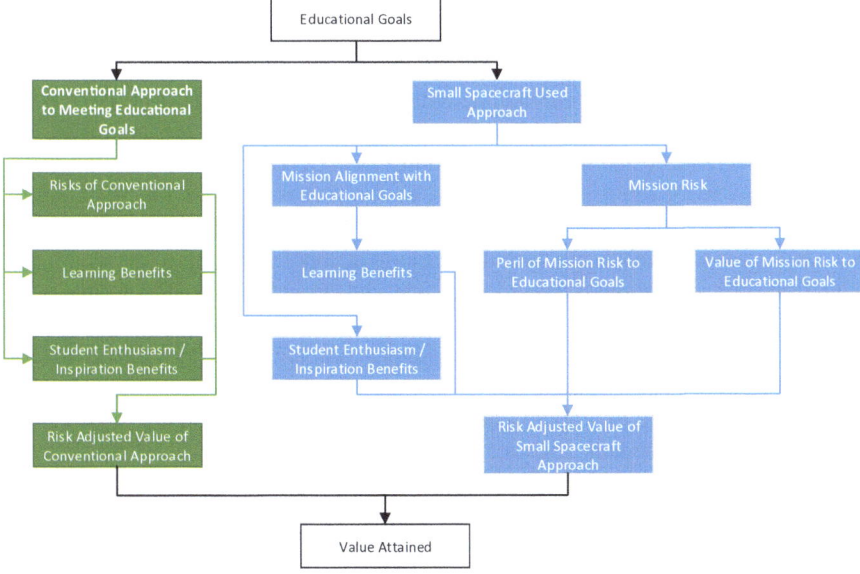

Fig. 6.1 Model for assessment of small satellite development use learning value

lesson allow a student to do in the future; what idea might it spark). In more practical terms learning value can be assessed in terms of the amount of resources consumed and the learning goals attained. For example, a one semester class of which one class period is spent on an activity should (based on a 16 week schedule, three classes per week) cover approximately 2.1 % of the material for the course. If a budget exists and there is an activity cost, it should draw upon this proportionately to the amount of material that is covered.

The value of student enthusiasm cannot be overstated. The actual calculation or quantification of this is difficult, despite its conceptual simplicity. The enthusiasm metric is the difference in the enthusiasm about the subject, class, and related material between the initial condition (prior to the learning experience) and the final condition. This can be assessed anecdotally via the level of interest and desire to participate shown; it can be assessed more formally via surveys or via looking for attributable changes in performance metrics (grades on exams and assignments, etc.). Increasing student interest has been shown to increase both knowledge acquisition and retention [29, 30].

The instructional approach utilizing the small satellite development learning technology is assessed similarly. The differences are now discussed. These include the curricular structure required to maximize learning, the value of involving risk in the learning experience, and a modified strategy for determining risk-adjusted value (which incorporates the aforementioned).

Relative to most in-class or course lab exercises, small satellite development is both expensive and time consuming. Expenses (which include the small satellite, the payload, launch and other expenses for operations) are significant. CubeSat development, thus, cannot be a replacement for a single lecture or small number of lectures and produce a strong learning value return on investment under the model that was discussed for conventional activities. The launch and on-orbit operations should serve as a culminating experience for other work. Under this approach, the project can provide value in terms of learning facts (from the success or failure of the decisions made) and skills (in terms of the processes used for preparation, launch, and operations), and it has strong experiential value. Problematically, this is not easy to assess. In fact, much of the learning occurs leading up to the flight; the actual on-orbit operations serve to reinforce and provide feedback on this prior learning and work.

Because of this, it is not appropriate to simply compare the hours of development and operations to a corresponding use of time (say, for example, consumed by lecture style instruction). The two approaches must be compared holistically. That is, the small satellite development relevant course components (design of the spacecraft, required research, testing, etc.) must be compared to corresponding components (or a corresponding use of time) under the non-small satellite-involved course approach.

In the assessment of the traditional style approach, risk was seen as an impairment to achieving learning objectives. Under the small satellite development approach it serves this role, but this is only part of the risk equation. The development, launch, and operations experiences bring with them significant risk: risk of design or fabrication issues, risk of damage to the spacecraft, experiments, or equipment contained inside it, and various other types of risks associated with the integration, testing, launch, and deployment phases. While several of these categories of risk are small, they provide an excellent opportunity for teaching students to think about experimental and implementation risks. Thus, risk must be assessed for both its benefit to and its impairment of educational value.

The impairment calculation was discussed previously and results in a sum of likelihood-of-occurrence adjusted magnitude-of-impact values. For each category of risk identified, however, students can develop mitigation-of-occurrence and contingency response plans. At a minimum, this teaches project management knowledge and skills (which may or may not align with learning objectives); however, this value may be further increased by tying risk response efforts back to learning objectives. For example, a related project might develop and use skills in a designated learning area to assess or develop a response plan for a prospective risk. The areas prospectively covered could include everything from engineering (developing a payload structure to mitigate property damage from occurring in a collision) to life sciences (how does a given condition affect a biological experiment or payload) to psychology (what impact could this type of failure have on participants and others) and law (what liability might exist and why).

The new risk model combines a probability-of-occurrence, magnitude-of-impact peril model with probability-of-occurrence, magnitude-of-benefit, and pre-attained benefit components. This model is shown in Fig. 6.2. Previous work [31] discussed the risk factors relevant to student-involved projects. While quantification of the level of occurrence of these risks (enabling projection and analytic risk analysis)

6.3 Matching Goals and Funding Sources

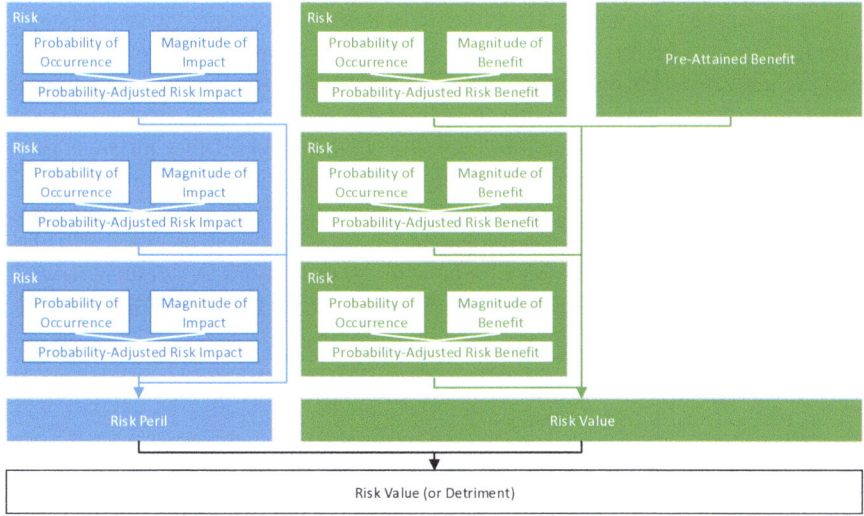

Fig. 6.2 Risk model

hasn't been performed (an example of ongoing work is Brumbaugh and Lightsey's [32] attempt to assess mission risk in the context of CubeSat missions), a qualitative assessment can be valuable from a formative perspective to facilitate risk identification and avoidance.

6.3 Matching Goals and Funding Sources

A critical component of goal selection is to ensure that the goals are fundable to a sufficient level to bring the development of a small satellite to fruition (or establish a longer-term program, if desired). For commercial endeavors, this may be a simple question of finding an internal or external customer who is willing to pay for spacecraft development and launch. An entity may also pay for its own development and launch in the hopes of generating revenue to recoup the development and launch expenses and generate a profit.

On the noncommercial side, the funding for development and launch may need to be considered separately. In the United States and ESA member states, two programs (the Educational Launch of Nanosatellites [33] and Fly Your Satellite [34], respectively) provide launches for small spacecraft without associated development funding. United Launch Alliance [35] has also offered a limited number of 1-U CubeSat spots on its rockets to US educational institutions, also without funding for spacecraft development. The University NanoSat Program [36], alternately, provides limited funding for development; however, developers may need to augment these funds with other funding to meet their mission goals.

Because of the potential for a low or no-cost launch, not coupled to development funding (or with insufficient funding to complete development), developers may need to carefully design their mission to meet both launch-funder and development-funder objectives. For some missions this may be easy, with complementary science and educational objectives being acceptable to both funders. Other missions may find themselves able to secure launch, but not development, funding or vice versa. Some educational institutions may choose to apply for an ELaNa, ULA, or Fly Your Satellite launch and self-fund development as an educational or extracurricular activity; however, this may not be affordable for all institutions. A careful review of mission objectives relative to prospective funding for both launch and development is, thus, called for.

6.4 Goal-Based Technique for Requirement and Constraint Decision Making

With program/mission objectives determined, the next step is to execute towards achieving these goals. This section describes the definition of requirements and constraints, based on mission goals and other factors. It then discusses how these identified characteristics can be used as a foundation for effective decision making.

6.4.1 Defining Requirements and Constraints

Requirements and constraints should be specific, quantifiable (where possible) statements that can be evaluated as being attained (or not). Requirements and constraints are generated from objectives as well as additional information.

6.4.1.1 Functional and Operational Requirements

Functional and operational requirements define capabilities that the system must have (functional) and be able to do (operational). There are two key considerations when generating requirements. The first is the mission objectives. Each objective should be decomposed into one or more requirements. The complete set of requirements associated with each objective should be sufficient to ensure that the objective is met, if all requirements are met. The second key consideration for requirement generation is the testability of the requirement. Peter Drucker famously noted the extreme difficulty of managing what cannot be measured [37]. Ensuring that your requirements are measurable eases management processes and avoids later confusion and disagreements.

6.4.1.2 Constraints

Constraints share many traits with requirements and could, generally, be reworded and presented as requirements. However, the separation is valuable for working purposes, as the two may originate from different sources. Constraints can be considered as being restrictive statements of what a project cannot do (while requirements are positive statements of what it should/must do). Constraints can relate to budget, schedule, safety considerations, legal considerations, ethical consideration, and such. A successful project must, thus, satisfy all requirements and not violate any constraints. Like requirements, constraints must be test-able and specific.

6.4.2 Creating and Selecting a Mission Concept

With the objectives, requirements, and constraints in place, a variety of brainstorming techniques can be used to identify approaches that may fulfill them. One approach that can be taken for concept generation and selection is based on the approach to conducting spontaneous creativity challenges applied by the Odyssey of the Mind organization.

This creative problem solving activity is designed to produce a large set of divergent answers within a short period of time. The approach also combines the benefits of the two previously described systems of idea generation. In the competition, participants are given 1 min to silently think and two to respond [38]. Participants, thus, benefit from generating ideas without interruption or having their direction of focus shaped by others involved in the process. The communal sharing, however, also provides the opportunity for stating ideas which 'piggyback' off of the ideas of other team members.

It is suggested that participants in the mission concept generation process be given a set amount of time to record as many possible approaches to satisfying the objectives and requirements as come to them. The ideas should be recorded as short conceptual statements and not developed any further. Once participants are done, the ideas can be shared, in a round-robin fashion, with others in the group. Participants should be encouraged to record and share any additional ideas that come to them during this process. No judgment should be made—all non-duplicative ideas should be recorded by the process leader.

A mission concept should provide a complete answer to how the mission will be conducted, albeit with a low level of specific detail. For orbital missions, it is recommended that the key questions to be answered include what data will be collected and what will be done with it, how data and commands will be transmitted to and from the craft, how the activities of the craft will be decided and controlled, and what the timeline of the mission is [39]. These questions are also very relevant for small spacecraft missions.

6.4.3 Using Objectives, Requirements, and Constraints for Decision Making

The mission objectives as well as the requirements and constraints defined from them serve as a key guide to most mission decision making. The mission objectives, in particular, determine what must be done (primary objective/objectives), what should be done if it doesn't impair or conflict with the primary objective (secondary objective/objectives), and what is part of the mission, but less important (tertiary objectives). These objectives are translated into requirements; however, this translation may be imperfect and requirements may need to be reevaluated relative to the objectives should issues arise. Absent this, the requirements and constraints (which represent standards that must be adhered to, or part, integration, or vendor limitations, and such) drive the day-to-day decision making. Each element of the spacecraft and each action taken must be able to be tied back to a requirement or constraint. Anything that cannot must be evaluated to determine if this indicates a fault with the requirements and constraints and, failing this, should not be included or performed.

6.5 Picking a Design Framework and the Level of Program Rigidity

A critical component of the design of any small spacecraft mission is the design approach that is utilized. Several common frameworks exist; however, these may be too complex for many missions. This section discusses existing design frameworks and adapts a lightweight framework (initially presented for use with high altitude ballooning in [1]) for use by small satellite programs.

6.5.1 Comparison of Design Frameworks

The goal of the simplified mission analysis and design (SimplMAD) process is twofold. First, it is designed to provide a framework that is right-sized to the design of most small spacecraft missions. The three-step framework can be scaled up by spending additional time and resources on various subcomponents. Alternately, by minimally covering each of the three design phases, a small mission can be designed in an amount of time commensurate with its scope.

Table 6.1 contrasts the SimplMAD model with the models presented in Space Mission Architecture and Design, 3rd Edition (SMAD 3) [39], Space Mission Engineering: the New SMAD (SMAD 4) [40], and Spacecraft Systems Engineering, 4th Edition (SSE 4) [41]. These texts form the basis of most university space mission design courses and the alignment of SimplMAD with these common frameworks makes it suitable for an introductory space mission design course by ensuring that the knowledge gained can be applied to the follow-on, more detailed courses.

6.5 Picking a Design Framework and the Level of Program Rigidity 87

Table 6.1 Correlation of mission engineering phases

SimplMAD	SMAD 3 [39]	SME-SMAD [40]	SSE 4 [41]
A. Defining objectives, requirements, and constraints	1. Definition of mission objectives	1. Define the broad objectives and constraints	A. Feasibility
		2. Define the principal players	
		3. Define the program timescale	
	2. Preliminary estimate of mission needs, requirements, and constraints	4. Define the quantitative needs, requirements, and constraints	
B. Concept and architecture development	3. Identifying alternative mission concepts	5. Define alternative mission concepts	
	4. Identifying alternative mission architectures	6. Define alternative mission architectures	
C. Drivers, requirements, analysis, and selection	5. Identifying system drivers	7. Define the likely system drivers and key requirements	B. Detailed definition
	6. Characterizing the mission architecture	8. Conduct performance assessments and system trades	
	7. Identification of critical requirements		
	8. Mission utility	9. Evaluate mission utility	
	9. Mission concept selection	10. Define the baseline mission concept and architecture	
		11. Revise the quantitative requirements and constraints	
		12. Iterative and explore other alternatives	
		13. Define system requirements	C/D. Design, development, manufacture, integration, and verification
		14. Allocate the requirements to system elements	E. Mission operations and data analysis

One approach that could be taken would be to begin with an introductory course that includes a complete small spacecraft project. This would be followed by a set of courses that cover each subsystem and payload design in greater detail. A capstone course, utilizing SMAD 3, SMAD 4, or SSE 4 could then complete this process. The use of SimplMAD for the introductory course allows students to comprehend the value of learning about the subsystems, without getting bogged down in detail.

6.5.2 Mission Analysis and Design

The SimplMAD is designed to be a lightweight approach that mirrors critical elements of the space mission design process. This three-phase approach begins with the definition of objectives, requirements, and constraints. From this, a mission concept and architecture are developed. The mission architecture is used to define and analyze critical drivers, which are used to create a final mission plan.

6.5.2.1 Defining Objectives, Requirements, and Constraints

Once a prospective mission is conceptualized, it must be formalized by identifying the pertinent objectives, needs, requirements, and constraints. Objectives are, quite simply, the goals that drive the creation of the solution. Goals should be broad in nature and specify what is desired to be accomplished—not how it should be accomplished.

The needs defined by the objectives flow through into the definition of requirements and constraints. Requirements are specific statements that a mission concept must achieve in order to successfully satisfy the objectives that have given rise to the requirement (requirements that cannot be tracked to an objective should be examined carefully and likely removed). Wertz and Larson propose that requirements should be broken down into two distinct categories: functional requirements, which define the desired performance characteristics, and operational requirements, which define system operation and user interaction [39].

Constraints are effectively negative requirements, which remove a part of the solution space from consideration. Constraints may be generated from objectives; however, they can also be born from economic and programmatic realities (such as the level of budget available). Requirements and constraints can be either quantitative or qualitative, but must be specific enough that compliance with them can be easily determined.

6.5.2.2 Concept and Architecture Development

The process of concept and architecture development is similar—albeit at different levels of detail. The mission concept is the first level where one determines how the mission will be conducted. Wertz and Larson proffer that at least four key questions

6.5 Picking a Design Framework and the Level of Program Rigidity

should also be answered: what data will be collected and how will it be provided to its users, how will various parts of the solution system talk to each other, how will the system be controlled, and what is the schedule of the mission project [39].

The architecture is even more detailed. At this level, however, the focus turns to trading various elements with each other to maximize mission performance in terms of the metrics defined by the objectives, requirements, and constraints. The mission concept, under the approach taken by Wertz and Reinert, forms one of the possibly tradable elements [39]. Table 6.2 describes the SimplMAD mission architecture elements.

6.5.2.3 Drivers, Requirements, Analysis, and Selection

The final steps of the mission design process involve performing analysis in support of final selection and making a final selection that defines all elements of the mission. This process begins by identifying drivers: the features of the mission that are controllable and have influence on key mission metrics including cost, schedule, and performance. Risk, while not a stated metric, is also a source for drivers, as it impacts the ability of a mission to deliver on the other metrics. Driver identification can be performed by starting with the key metrics and reviewing each controllable mission element to determine whether changing it impacts the metric.

The trade analysis process seeks to maximize the mission utility via selecting the best set of mission requirements. Utility analysis requires that each metric be quantifiable (even if this quantification is arbitrary and only done for the purpose of this analysis) and that the relative importance of the metrics be defined via the assignment of coefficients. Each possible solution then has its score calculated and the one with the highest utility value wins.

Table 6.2 Mission architecture elements

Element	Description
Mission concept	Approach that is taken to the mission
Subject	The target of the mission: what is being imaged, sensed, or affected by the mission
Payload and subsystem elements	Various integral components that together provide the capabilities to perform whatever actions the mission must take. Only critical elements should be identified at this point
Spacecraft bus	The spacecraft bus will house, provide a structure for, power, possibly deploy, and protect other components of the spacecraft system
Communications approach	Will the mission involve communication with the ground? Just one-way position transmission? One-way data transmission? Two-way data transmission and control?
Operations approach	Will the mission require a control station? Only monitoring? What level of staffing (24×7, limited, etc.) will be required?

Once iteration does not seem to be having a meaningful impact in increasing solution utility, it is time to pick a mission solution. This process starts with the solution that has the highest utility. The solution must then be evaluated to ensure that it meets all requirements and constraints. If so, it is selected; if not, further refinement may be required or an alternate solution must be considered.

6.5.3 Defining the Mission Architecture

A mission architecture is born from a mission concept and enumerates a set of mission characteristics that flow from the concept. The creation of several concepts and architectures (possibly including multiple architectures born from a single concept) is desirable to ensure that the mission solution space is well explored before an architecture is selected. For near-space missions, architecture elements include the subject, payload elements, and bus, target altitude and mission duration, ground systems, and communications approach.

6.5.3.1 Subject

The subject is the reason for conducting the mission. It is the target of the mission's investigations. This would include a remote sensing target or an onboard plant or animal, whose exposure to near-space conditions was being observed.

6.5.3.2 Payload Elements and Bus

For a basic mission, the payload elements include the transmitter that is used for recovery, any onboard instruments, and/or any onboard experiments. A basic small spacecraft mission may use a shared bus provided by a third party; however, more complex missions typically will consist of a bus (core, potentially common across satellites, infrastructure of a satellite) and a variety of interconnected subsystems.

6.5.3.3 Target Altitude, Orbit, and Estimated Mission Duration

The target altitude, exact orbit selected, and the estimated mission duration (based on the altitude, solar pressure, and other factors [40]) are key mission considerations. These architecture elements are born from mission requirements related to the subject of the study and the duration of time that is required in orbit.

6.5.3.4 Ground Systems

At a minimum, it is required that telemetry be received in order to determine where the spacecraft is and what point it is at (ascent, peak altitude, descent) during its flight. The specific communications plan for the mission may dictate additional requirements for ground stations, if extended telemetry is being transmitted or commands will be sent to the spacecraft.

6.5.3.5 Communications Approach

The communications approach determines when the spacecraft payload will be communicated with and what will be communicated. Some small spacecraft missions may choose to support only one-way communications, providing only telemetry and data downlink. Most spacecraft will also receive commands (and possibly other transmissions) from the ground and take corresponding actions.

The communications approach that is selected will also have a significant bearing on the autonomy of the craft (or conversely, a decision to operate autonomously or not may drive the communications approach). Craft that do not support two-way communications must operate independent of any ground support throughout the mission.

6.5.4 Driver Identification

The refinement of a mission architecture is performed by identifying the elements that affect it and determining the impact of trades (changes that may add benefit in one area and reduce the benefit in another). Drivers are the mission elements that impact cost, schedule, and other key metrics.

Driver identification can be performed by starting with the key metrics and reviewing each controllable mission element to determine whether changing it impacts the metric. If it does, the element is a driver for the metric. Some elements may be identified as having an impact only in conjunction with another element.

6.5.5 Requirements, Analysis, and Selection

With the drivers identified, the key requirements — those which have the most impact in determining the mission's performance in terms of metrics (e.g., cost, schedule, performance) — can be identified. Key requirements can be identified by reviewing the identified drivers, identifying what requirements influence them, and how significantly. Key requirements are the requirements that have a significant impact on one or more drivers — or a more minimal impact on numerous drivers. The identification of key requirements is a critical part of solution selection as the key requirements are the focus of the requirements trade analysis process. The trade analysis process will, logically, focus only on the key requirements that have been deemed tradable, previously.

The trade analysis process seeks to maximize the mission utility via selecting the best architecture that fulfills all mission requirements. Utility analysis requires that each metric be quantifiable (even if this quantification is arbitrary and only done for the purpose of this analysis) and that the relative importance of the metrics be defined via the assignment of coefficients. Each possible solution then has its score calculated and the one with the highest utility value wins. The identification of key metrics constrains the search space (the number of combinations that should be considered) by allowing the process to focus on only the most important possible trades. Practically, the process is somewhat more complicated than this as the analyst may identify new possible solutions upon seeing what elements have the most impact and what prospective solutions perform the best. Given this, an iterative process will likely occur with possible solutions refined and compared several times.

Once iteration does not seem to be having a meaningful impact in increasing solution utility, it is time to pick a mission solution. This process starts with the solution that has the highest utility. The solution must then be evaluated to ensure that it meets all requirements and constraints. Its risk must be evaluated to ensure that it is acceptable. If any of the above validations fail, the solution may need to be further retooled (and compared to others, if its utility value has changed). The result of this final step is to choose a mission solution and make a go/no-go decision as to whether to proceed with the mission at all.

6.6 Planning for Program Longevity: Technical and Logistical Considerations

A near-space mission consists of several distinct high level phases. These phases are applicable to any type of mission; however, for smaller missions, they may be conducted informally, with limited process and procedure implementation. The mission starts with the conceptualization phase. It then proceeds into the design phase where objectives are identified and iteratively refined and the various craft components and their interoperation are identified. Next, the mission moves into the development phase, where systems are built, tested, and refined. Once craft construction is complete, the mission proceeds to the launch and operations phase where the craft is sent up, data is returned, and operational decisions are made. Finally, when the mission is done, the conclusion phase results in the documentation of mission activities.

6.6.1 Conceptualization

The conceptualization phase serves, primarily to the answer to the question of why: why undertake the activities? For larger missions, this concept may come from a program objective, sponsor mandate, or competitive proposal evaluation process. Smaller

6.6 Planning for Program Longevity: Technical and Logistical Considerations 93

missions may be conceptualized by an individual who has appropriate authority and has decided to pursue an identified goal. The conceptualization phase concludes when a concept is selected (formally or informally) and design activities begin.

6.6.2 Design

The design phase should start with the identification of mission objectives and a consideration of whether a small satellite mission is an appropriate way to achieve these objectives. The comparative utility (benefit and associated cost) of other possible approaches should be considered. If a decision to proceed with a small satellite mission is made, at a minimum, this phase must determine what will be launched (e.g., specific details including design details for any components that will be fabricated), how it will be launched (e.g., primary rocket payload, secondary rocket payload, International Space Station deployment), and any constraints related to when it will be launched (e.g., it must be launched when Venus can be seen in the sky).

Payload design will flow from the answers to these high level questions. An iterative process of refining the near-spacecraft design from a high level concept to actual identified parts and integration methods will result in the specifications required to begin the development phase.

6.6.3 Development

The development phase includes not only fabricating, integrating and testing components but also an iterative process of ensuring that the components work together as a system. Components should be fabricated or procured and tested individually (called unit testing) and then assembled and tested together (called integration or system level testing). For larger projects, clusters of components (called assemblies) can and should be tested before being incorporated as a piece of a larger cluster. The development phase concludes when all requirements are met (or deviation is documented and approved) and this adherence has been affirmed via successful unit and system level testing.

6.6.4 Launch and Operations

At present, the process of arranging a launch for a small spacecraft has a very well-defined process. This may change, at some point, when small (single small satellite or small satellite-as-primary launch vehicles become prevalent). Typically, a small spacecraft is a secondary (or tertiary) payload on the launch vehicle and, thus, may be required to be assessed to a similar level of safety as the primary payload. It is also critical to demonstrate that the spacecraft cannot harm the primary payload. In many cases, operating 'turn-on' delays and similar may be required to facilitate gaining separation for safety.

The operations phase begins at launch. Depending on the mission specifics, this phase may consist of communication with and commanding the spacecraft and/or analyzing data and adapting the mission based on what has been discovered. In any event, the operations phase is the key time during the mission where engineering work is tested, objectives are achieved, and relevant data is collected. The operations phase concludes when the mission activities are terminated (e.g., when all objectives are complete, the spacecraft is unexpectedly lost or damaged and/or the spacecraft reenters the atmosphere).

6.6.5 Closeout

Any project or mission requires a period of time following its main activities to clean up. For academic and scientific missions that are part of an ongoing program, this may be as simple as documenting the missions' success (or failure), assessing the consumption of supplies, and returning reusable hardware to appropriate storage locations in preparation for future missions. Missions with scientific goals may require data reduction and reporting to be performed. Larger missions (particularly those that are not part of a continuing program) may need to follow a more rigorous conclusion process. The conclusion phase should wrap up the loose ends of the project.

6.7 Processes for Mission Management

Small spacecraft missions, like any project, require strong management to be successful. In an academic setting, this management need allows the expansion of involved students to include those who may be pursuing business or public administration degrees (in additional to the traditional engineering focus) and desire an experience in project management. Irrespective of student involvement, however, maintaining control of the mission is the only way to ensure a successful result. Important considerations include project/mission management, implementing appropriate systems and processes, and assurance activities.

6.7.1 Project/Mission Management

Planning, as an iterative process, can expand to fill whatever time is available to it. In many cases, this time expansion occurs without any benefit in terms of planning quality or outcomes. As such, it is critical to properly manage the planning process.

In fact, the first step in planning management should be to make a plan for the planning process. Specifically, this plan should identify the required outcomes, verifiable milestones, and the artifacts (documents) to be produced.

The defined outcomes should include both technical (problem solving/design) and team interaction goals. Just as the plan itself should include verifiable milestones to allow project sponsors and others to assure that the mission is proceeding as planned, the plan for the planning process should also include milestones.

It is critical to define what specific artifacts should be produced during each phase and any format constraints which are applicable. This plan should also identify target completion dates and include a management time reserve to accommodate the invariable slippage that will occur when a technical problem is discovered.

It may seem, at first glance, like the planning process can be ignored or dramatically simplified for small missions—like many academic missions. However, in some ways these small and academic missions require the planning process to a greater extent than large ones. Small and academic missions will likely utilize the services of individuals who have alternate full-time commitments. These individuals will have various levels of commitment, which may vary from week to week, due to other pressures, which they face [31]. By defining what is required from each member during the planning process, the leader is ensuring that a clear understanding is held by all participants—and creating a document that can be used to later remind individuals of the commitments that they have made and the impact that failing to meet them will have on the large group.

6.7.2 Systems and Processes

Any effective management methodology must employ systems and processes to control and document the various management and managed activities that are performed during a project or mission. Small spacecraft missions are no exception to this rule. Systems and processes should be employed starting from project initiation to track objective, concept and requirements generation, and any changes that are made to these and other key project elements.

A management system should be selected. One management system that is very well suited to small projects and also scales reasonably well is management by exception. The fundamental notion of management by exception is that the manager determines what an acceptable range of performance is (e.g., an upper and lower bound of time that a task should take). Processes that perform within the designated acceptable range are not reviewed (except, perhaps as part of an overall process audit), allowing the majority of the manager's time to be spent on tracking areas that are significantly overperforming or underperforming expectations.

Project deliverables must also be retained and tracked. The objects that need to be retained and tracked fall into two primary categories: artifacts and deliverables. Artifacts are any document (or similar) associated with the project that is not a defined outcome of the project (e.g., management documents, change tracking logs). Deliverables are, quite simply, anything that must be provided to a stakeholder as a part of completing the project's requirements.

Both artifacts and deliverables must be retained and tracked appropriately. However, the process that is implemented differs somewhat depending on whether they are physical objects or electronic documents (including software).

Changes that impact mission objectives, requirements, or constraints are particularly problematic, after decisions that rely on these foundational elements are made. Given this, most projects devote substantial efforts to the management of changes that impact these areas. By tracking and documenting these items one can ensure that the change's impact is properly propagated throughout the project. Tracking the changes also allows identification of what various cost and schedule overruns are attributable to.

6.7.3 Assurance

There is little point to having objectives, requirements, or constraints if action is not taken to ensure that these elements are met by project activities. Assurance activities ensure that defined high level parameters are met by lower level design and development activities. They also ensure that artifacts and deliverables meet the specifications required of them.

Requirements mapping is a technique that can be used to ensure that various high level elements (e.g., objectives, requirements, and constraints) are implemented in lower level design documents. With requirements mapping, the performing team member is required to determine and document how each high level element is implemented in the area being reviewed.

Even with the best of intentions, mistakes do happen. Quality management mitigates these risks by identifying areas where high quality is required and defining assurance activities to validate that this quality exists. Quality management can be conducted in two ways. One approach to quality management is to design it into a production or operations system. A second approach to quality management is validation based. In many cases, this approach is called for due to difficulties incorporating quality directly into a process or the high cost of a quality-integrated process failing.

6.8 Conclusion

This chapter has focused on the logistical and technical aspects of the formation of a small spacecraft program. It has considered the reasons why a program may be initiated, from an educational, scientific, engineering, or other perspective. It has then gone on to look at how mission and program objectives can be generated. Next, the definition of requirements and constraints was discussed. Then, selection techniques for choosing a design framework were presented. Finally, management techniques and longevity strategies were discussed.

While each program, its goals and operating environment, challenges, and design process will be necessarily different, it is hoped that the foregoing will help in the development of processes that can aid a nascent mission in meeting its goals and, perhaps, transforming into a longer-term program. Key to this is the identification of suitable goals and execution, using a design framework and proper management techniques, towards meeting them.

References

1. Straub, J., and R. Fevig. 2012. Formalizing mission analysis and design techniques for high altitude ballooning. Presented at Proceedings of the 3rd Annual High Altitude Conference.
2. Straub, J. 2015. Evaluation of high-altitude balloons as a learning technology. *International Journal of Learning Technology* 10: 94–110.
3. Chirayath, V., and B. Mahlstedt. 2012. HiMARC 3D-high-speed, multispectral, adaptive resolution stereographic CubeSat imaging constellation. Presented at Proceedings of the AIAA/USU 2012 Small Satellite Conference.
4. Bashevkin, E., J. Kenahan, B. Manning, B. Mahlstedt, and A. Kalman. 2012. A novel hemispherical anti-twist tracking system (HATTS) for CubeSats. In *Proceedings of the 26th AIAA/USU Conference on Small Satellites*.
5. Muylaert, J., R. Reinhard, C. Asma, J. Buchlin, P. Rambaud, and M. Vetrano. 2009. QB50: An international network of 50 CubeSats for multi-point, in-situ measurements in the lower thermosphere and for re-entry research. Presented at ESA Atmospheric Science Conference. Barcelona, Spain.
6. Deepak, R.A., and R.J. Twiggs. 2012. Thinking out of the box: Space science beyond the CubeSat. *Journal of Small Satellites* 1(1): 3–7.
7. Swartwout, M. 2004. University-class satellites: From marginal utility to 'disruptive' research platforms. Presented at 18th Annual AIAA/USU Conference on Small Satellites. 11pp.
8. Crowder, R.M., and K.-P. Zauner. 2013. A project-based biologically-inspired robotics module. *IEEE Transactions on Education* 56(1): 82–87. doi:10.1109/TE.2012.2215862.
9. Redkar, S. 2012. Teaching advanced vehicle dynamics using a project based learning (PBL) approach. *Journal of STEM Education: Innovations and Research* 13(3): 17–29.
10. Nordlie, J., and R. Fevig. 2011. Blending research and teaching through high-altitude balloon projects. In *Proceedings of the 2nd Annual Academic High Altitude Conference*. Ames, IA.
11. Brodeur, D.R., P.W. Young, and K.B. Blair. 2002. Problem-based learning in aerospace engineering education. Presented at Proceedings of the 2002 American Society for Engineering Education Annual Conference and Exposition.
12. Fevig, R., J. Casler, and J. Straub. 2012. Blending research and teaching through near-earth asteroid resource assessment. Presented at Space Resources Roundtable and Planetary & Terrestrial Mining Sciences Symposium.
13. Straub, J., J. Berk, A. Nervold, and D. Whalen. 2013. OpenOrbiter: An interdisciplinary, student run space program. *Advances in Education* 2: 4–10.
14. Fevig, R., J. Casler, J. Straub, R. Lilko, and C. Church. 2012. Blending research and teaching through small spacecraft development projects. In *2012 European Cubesat Symposium*. Brussels, Belgium.
15. Barron, B.J., D.L. Schwartz, N.J. Vye, A. Moore, A. Petrosino, L. Zech, and J.D. Bransford. 1998. Doing with understanding: Lessons from research on problem-and project-based learning. *Journal of the Learning Sciences* 7(3-4): 271–311.

16. Jackson, K., R. Fevig, and S. Seelan. 2012. North dakota state-wide high altitude balloon student payload competition. In *The Proceedings of the 3rd Annual Academic High Altitude Conference*. Memphis, TN.
17. Straub, J., and D. Whalen. 2014. Evaluation of the educational impact of participation time in a small spacecraft development program. *Education Sciences* 4(1): 141–154.
18. Simons, L., L. Fehr, N. Blank, H. Connell, D. Georganas, D. Fernandez, and V. Peterson. 2012. Lessons learned from experiential learning: What do students learn from a practicum/internship? *International Journal of Teaching and Learning in Higher Education* 24(3): 325–334.
19. Ayob, A., R.A. Majid, A. Hussain, and M.M. Mustaffa. 2012. Creativity enhancement through experiential learning. *Advances in Natural and Applied Science* 6(2): 94–99.
20. Doppelt, Y. 2003. Implementation and assessment of project-based learning in a flexible environment. *International Journal of Technology and Design Education* 13(3): 255–272.
21. Edwards, A., S.M. Jones, E. Wapstra, and A.M. Richardson. 2012. Engaging students through authentic research experiences. Presented at Proceedings of the Australian Conference on Science and Mathematics Education (Formerly UniServe Science Conference).
22. Bauerle, T.L., and T.D. Park. 2012. Experiential learning enhances student knowledge retention in the plant sciences. *HortTechnology* 22(5): 715–718.
23. Breiter, D., C. Cargill, and S. Fried-Kline. 2013. An industry view of experiential learning. *Hospitality Review* 13(1): 8.
24. Mills, J.E., and D.F. Treagust. 2003. Engineering education—Is problem-based or project-based learning the answer? *Australasian Journal of Engineering Education* 3(2): 2–16.
25. Straub, J., J. Berk, A. Nervold, C. Korvald, and D. Torgerson. 2013. Application of collaborative autonomous control and the open prototype for educational NanoSats framework to enable orbital capabilities for developing nations. In *Proceedings of the 64th International Astronautical Congress*. Beijing, China.
26. Straub, J. 2014. Extending the orbital services model beyond computing, communications and sensing. Presented at Proceedings of the 2014 IEEE Aerospace Conference.
27. Gill, E., P. Sundaramoorthy, J. Bouwmeester, B. Zandbergen, and R. Reinhard. 2013. Formation flying within a constellation of nano-satellites: The QB50 mission. *Acta Astronautica* 82(1): 110–117.
28. Straub, J., A. Mohammad, J. Berk, and A.K. Nervold. 2013. Above the cloud computing: Applying cloud computing principles to create an orbital services model. Presented at SPIE Defense, Security, and Sensing.
29. Lau, L.K. 2003. Institutional factors affecting student retention. *Education-Indianapolis then Chula Vista-* 124(1): 126–136.
30. Bean, J., and S.B. Eaton. 2001. The psychology underlying successful retention practices. *Journal of College Student Retention* 3(1): 73–89.
31. Straub, J., R. Fevig, J. Casler, and O. Yadav. 2013. Risk analysis & management in student-centered spacecraft development projects. In *Proceedings of the 2013 Reliability and Maintainability Symposium*. Orlando, FL.
32. Brumbaugh, K., and E.G. Lightsey. 2014. CubeSat Mission Risk Survey—A call for response from the small satellite community. *Journal of Small Satellites* 2: 83–84.
33. Skrobot, G. 2012. ELaNa—"Making it happen!" In *The 9th Annual CubeSat Developers' Workshop*.
34. 13 February 2013. *Call for Proposals: Fly Your Satellite!* http://www.esa.int/Education/Call_for_Proposals_Fly_Your_Satellite.
35. April 2016. *ULA STEM CubeSat Program 2016*.
36. Hunyadi, G., J. Ganley, A. Peffer, and M. Kumashiro. 2004. The university nanosat program: An adaptable, responsive and realistic capability demonstration vehicle. Presented at 2004 IEEE Aerospace Conference Proceedings.

References

37. MacLennan, A. 2010. *Strategy Execution: Translating Strategy into Action in Complex Organizations*. New York: Routledge.
38. Creative Competitions, Inc. 2006. *Animal Rhymes*. http://www.odysseyofthemind.com/practice/default_problem_details.php?problem_ID=1&group_ID=2.
39. Larson, W.J., and J.R. Wertz. 1999. *Space Mission Analysis and Design*, 3rd ed. Hawthorne, CA: Microcosm Press.
40. Wertz, J.R., D.F. Everett, and J.J. Puschell. 2011. *Space Mission Engineering: The New SMAD*. Hawthorne, CA: Microcosm Press.
41. Fortescue, P., G. Swinerd, and J. Stark. 2011. *Spacecraft Systems Engineering*, 4th ed. West Sussex: Wiley.

Chapter 7
Student Involvement and Risk

7.1 Introduction

Student involvement in research and other projects is common at universities around the world. Through internships, part-time work, and other mechanisms, students also perform limited work for commercial, governmental, and other employers. Despite the prevalence of student involvement in the development of key technologies and their performance of numerous duties, the management literature contains little consideration of the specific risk elements introduced by student workers. Inexperienced workers (including students, interns, and junior employees) have particular characteristics that may create new risk sources and alter the likelihood and magnitude of typical risks.

An understanding of the impact of using student and other inexperienced student workers is particularly important in the case of aerospace projects due to the low defect tolerance, inaccessibility, and criticality of many projects. Small spacecraft, for example, are commonly integrated as secondary payloads on rockets carrying other orders-of-magnitude more expensive hardware. They must meet the same (or perhaps even more stringent) integration standards as the primary payload. Some small spacecraft have also been launched via the International Space Station, necessitating their compliance with human safety standards. Once they are in orbit, they are also on their own, with no practical servicing capability. Design and implementation failures can, thus, cause a spacecraft to fail integration testing and not get launched, to fail subsequent to integration and damage expensive equipment, or pose a threat to astronauts or fail on orbit, impairing mission performance. The training and research provided by these efforts is integral to developing new technologies as well as training the next generation of aerospace professionals. Given this, a better understanding of the risks posed by student and inexperienced staff involvement is necessary.

This chapter is based on and revises the paper "Extending the Student Qualitative Undertaking Involvement Risk Model" [1].

This chapter presents an enhanced model that augments the base Student Qualitative Undertaking Involvement Risk Model (SQUIRM) framework with root cause analysis, resulting in a more detailed consideration of student status on typical (nonstudent) risk factors. The use of this model can provide a more robust evaluation of the impact of student participation, as compared to the base model. However, it is not a panacea, and prospective trade-offs between the use of the two approaches are discussed. Second, it begins the process of quantifying the SQUIRM and extended SQUIRM frameworks, discussing how the models can be used in order to assess risks (considering likelihood, impact, and the mitigation techniques employed) on a single project basis or across multiple projects. Third, it presents a value model for evaluating the participation of student (and other inexperienced) workers. This model facilitates the determination of the value proposition of using this type of staff, which can be compared to increased risks and other associated costs. Finally, the differences between types of inexperienced workers are briefly discussed, before the conclusion.

7.2 Background

This section provides an overview of areas that the current work benefits from a wealth of prior work in. Despite a growing contemporary interest, the tasking of trainee or inexperienced workers to real-world projects is certainly not a new phenomenon. Apprenticeship style training has been used throughout history [2, 3]. Modern approaches, however, combine formal and experiential techniques. One relevant technique is project-based learning. In the remainder of this section, first the benefits of project-based learning are discussed. Next, prior work, regarding assessment of the value of students to faculty efforts, is briefly considered. Finally, a brief discussion about risk perception is presented.

7.2.1 Project-Based Learning

With project-based learning (PBL), students are involved in hands-on projects that could be developed specifically for a course or which might feature student involvement in faculty research or other real-world projects. PBL has been shown to be an effective instructional tool at all levels of education: from collegiate graduate level to primary school level [4–9]. It has also been demonstrated across a wide variety of subject disciplines, including project management [10], psychology [11], physics [12], computer science [13, 14], mathematics [15], engineering entrepreneurship [16] and aerospace [17, 18], computer [19], electrical [20, 21], and mechanical [22, 23] engineering.

In addition to teaching subject-specific skills, PBL projects can teach students how to work with those outside their specific discipline, as is required in the vast majority [24] of workplaces. Gaining a shared prior knowledge base (such as through PBL techniques) can improve team efficiency [24]. Workers with interdisciplinary skills are in demand [25]; PBL also provides students with an opportunity to learn "soft" skills which are required for workplace success [26].

7.2 Background

PBL has also been shown to have beneficial impact on student motivation [27], self-image and creativity [28], and material retention [29]. Field-based/realistic-environment PBL has been shown to increase students' understanding of course materials [30]. Nagda et al. [31] show that one type of PBL, research participation, can also improve student retention, particularly for at-risk students. The benefits of PBL to student placement, after graduation, have been demonstrated by Hotaling et al. [32] and Fasse et al. [33]. Gilmore [34] even argues that techniques such as PBL, for teaching STEM disciplines, are critical to national prosperity.

In aerospace engineering and related disciplines, many students are gaining practical experience working on small spacecraft and high altitude ballooning projects. The SQUIRM framework [35] was created, initially, to assess the risks applicable to student involvement in a small spacecraft project; however, it is useful for many applications beyond this. The utility of PBL for teaching aerospace engineering [4, 36], software development for aerospace applications [37], and providing other benefits [38, 39] has been demonstrated. CubeSat projects have been demonstrated to be an effective pedagogical approach [40–42].

The level of the aforementioned benefits, Zydney et al. [43] proffer, increases with the duration of participation. However, not all students reach these higher levels of benefit, while numerous reasons for premature termination of student participation in a research project exist, manifestation of the risk factors discussed in a subsequent section may explain some of the incomplete experiences.

7.2.2 Value of Student Involvement to Faculty Research

If student involvement's benefit was solely student education, the need to characterize and mitigate risks would be dramatically reduced. The impact of a student/inexperience-specific or general risk factor's occurrence can have impact to the student participant's success; it can also have a pronounced effect on the project as well. While students may gain (possibly even enhanced) benefit from risk actualization, the project stands to suffer. To characterize the magnitude of impact, it is important to consider faculty perceptions of student involvement on research projects. Zydney et al. [44] proffer that faculty see students' participation as valuable, with over half of them indicating that students' contribution to their work was "important" or "very important." Thus, the failure of a student to make progress is a risk that may be comparable to causing damage or other types of impact on prior work. While student participation is valuable to faculties, it appears that project completion may be less important to students, as Prince et al. [45] demonstrated a lack of correlation between the research productivity level of faculty and students' educational benefits.

7.2.3 Risk Perception

One reason that student workers may be more risk occurrence prone is a failure to properly assess risk likelihood and impact. However, despite a significant correlation between youth and inexperience, it is important to note the potentially confounding impact of risk perception. Because of this, there may be a performance difference between younger and older individuals with similar experience levels in a field. A full exploration of the topic of risk perception is far beyond the scope of this discussion; however, reviews of areas of this topic are readily available. Botterill and Mazur [46] provide a general overview of the topic, while Slovic et al. [47] consider the value of studying it. Boholm [48] reviews and compares risk perception research over a 20-year period and Mitchell [49] considers risk perception and risk reduction in the context of an organization.

The crux of the risk perception problem is that younger individuals may fail to appreciate the applicability of risk to them and its impact [50]. This has been documented across multiple areas, including driving [51], sexual [52], and other "health-threatening" [53] behaviors. Steinberg [54] attributes the greater risk-taking tolerance of youth to "age differences in psychological factors that influence self-regulation." Thus, age may confound the experience/risk correlation and intensify certain risk factors when both young age and inexperience are the case. Given this, traditional-age undergraduates may have a higher propensity to fail to see how their actions, behaviors, or inaction may create risks, or the impact that these risks may have on them or others.

Risk perception, however, is not only affected by age. Correlation has been shown with gender [55], culture [56], and other factors [57, 58]. The impact of education in correcting risk perceptions has been demonstrated by [59]. Weber and Milliman's [60] work suggests that "risk preference" may be a stable aspect of an individual's personality, highlighting the importance of risk perception on the acceptance or rejection of the risk in a given circumstance. Renn [61] discusses the importance of risk perception in relation to the management of risks.

7.3 The Student Qualitative Undertaking Involvement Risk Model

The following sections, reprinted with minimal modification from [35], provide an overview of the risk categories of the SQUIRM framework (which is depicted in Fig. 7.1). First, technical, schedule, and other standard risks will be discussed. Then, the risks posed by student worker involvement will be considered.

7.3.1 Technical, Schedule and Other Standard Risks

Every project, including those involving students, must deal with numerous possible risk factors. Project managers attempt to control many of these risk factors, assume others, and they are, ultimately, forced to ignore a large set of risks that they

7.3 The Student Qualitative Undertaking Involvement Risk Model

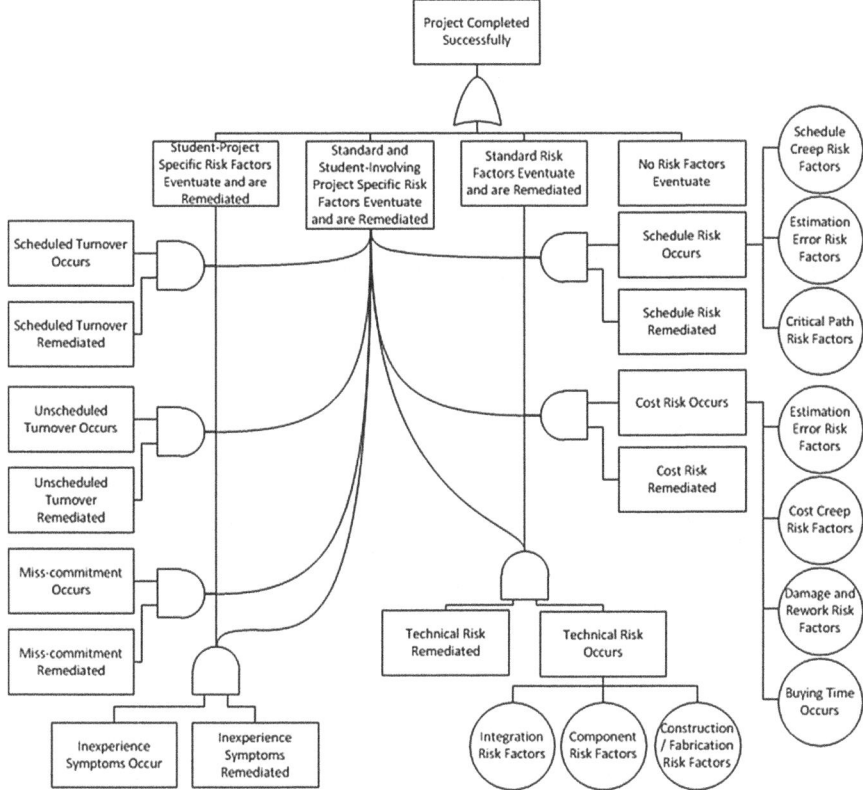

Fig. 7.1 SQUIRM model diagram [35]

have no insight into or control over. Numerous standard risks are well documented in the literature and will not be reviewed in detail here. The impact of student participation on these standard risks is considered. For each risk factor, a brief description of its nature is provided. This is followed by a discussion of how the risk factor is influenced by or may influence student project involvement.

7.3.2 Technical Risk

The technical risk category comprises the set of risks that could result from a failure of hardware and software or its integration and operations to perform as required to meet project's objectives. Three aspects are considered: construction/fabrication of assemblies, failure of purchased components, and their integration.

7.3.2.1 Construction/Fabrication

Construction and fabrication risks are inherent to any manufacturing process. Quality control processes, including those designed to prevent defects as well as those to detect and remediate defects, are generally included to mitigate these risks. In a student project, which generally doesn't involve mass production, one is confronted with two primary risks. First, standards-based quality control may be cost prohibitive to implement. Second, students who lack knowledge and understanding of the characteristics of the product may be poorly equipped to detect and evaluate the significance of errors.

7.3.2.2 Component

Components obtained from suppliers will occasionally be defective, either due to manufacturing or due to shipping issues. Production processes generally incorporate an acceptance testing procedure or supplier process validation procedure. A student-involved project, generally, suffers from two risk factors regarding components. First, the limited production (in many cases, producing only a single or small quantity of units) precludes the implementation of a standard quality process. Second, student inexperience may result in a failure to properly design acceptance tests or to detect latent issues.

7.3.2.3 Integration

The process of combining components together introduces risks due to design and implementation failures. Design failures may result in a system, which, regardless of how well it is assembled, cannot perform the desired task. Implementation issues may result in degraded performance, nonoperation, or failure after a period of time operating. Student designers and workers generally have traits that significantly increase the probability of these risks happening. Having an incomplete or largely untested understanding of the design process or specific design elements may result in wholly unworkable designs or designs with latent and hard-to-detect flaws. Limited time and resources will generally result in a comparatively lower level of testing being conducted. The fact that this testing will likely be performed by inexperienced (student) testers further exacerbates the problem. Even if a perfect design is produced, inexperience in the techniques required for construction may result in subpar construction, component attachment and solder connection issues, and so forth. These may cause the assembly to not work initially or to be prone to failure.

7.3.3 Schedule Risk

Every project faces the possibility that its schedule will not be met. External factors, such as the unavailability of key components, and internal factors, such as staff absences or equipment failure, may result in delays. When these delays impact the critical path,

7.3 The Student Qualitative Undertaking Involvement Risk Model

the project schedule is impaired. Key areas of consideration for projects involving students include schedule estimation error, critical path risks, and schedule creep.

7.3.3.1 Schedule Estimation Error

Estimation error occurs when the time projected for task completion is different from actual task completion. A certain amount of error is to be expected; however, when tasks are consistently taking longer than projected, the project's schedule is at risk. Estimation error is common, even for experienced estimators. Students, who do not have significant experience, may fail to consider anything other than the best-case scenario. Alternately, they may not completely understand the process that they are estimating and, thus, omit the time required for overlooked process components. Either of these may result in (possibly dramatic) underestimation. On the other hand, students may be overwhelmed and wildly overestimate (so as to avoid the pitfalls of underestimation). This is, however, problematic, as it may result in the project's momentum being lost, if materials, tools, or staff for subsequent phases are not available when a previous phase is completed early.

7.3.3.2 Critical Path Risk

Critical path risk is a set of risk factors that impact the chain of tasks, which, in succession, take the longest amount of time. As the project is not complete until all of these tasks are done, anything that elongates the schedule of a task on the critical path (or another task, which becomes a critical path task due to schedule overrun) affects the project's overall schedule. Critical path risk can be created by factors that are both external and internal to the project. External factors may include impairment to the availability of supplies, unavailability of key equipment at the needed time, changes in laws or regulations, and many other factors. Internal factors, however, are the primary area where projects with student involvement differ from conventional projects. Internal issues that may be exacerbated by student involvement include staff availability issues, delays caused by quality failures (and, thus, the need to repair or recreate the improperly produced items), and delays caused by poor scheduling. Staff availability and quality issues are discussed in other areas of the model. Poor scheduling may be the result of a failure to identify precursor and successor tasks due to failing to identify required task inputs and outputs or, more simply, error in the actual creation of the schedule. Either of these can easily occur when a schedule is produced by an inexperienced scheduler.

7.3.3.3 Schedule Creep

Schedule creep is the schedule component of scope creep. Scope creep occurs when changes or documentation issues result in a more robust product being produced than the one called for by planning. The involvement of students, who are generally eager to please and may not understand the impact of accepting changes (or not understand that they are implicitly accepting a change), increases the risk of schedule creep. The fact that most academic projects are run by professors who are trained as researchers—not project managers—and may have limited documentation further exacerbates this risk.

7.3.4 Cost Risk

With tight budgets and long-duration funding cycles, cost overrun is a significant risk to student-involved projects. Cost overruns can lead to reduced deliverable utility and/or quality. If severe enough (and supplemental funding cannot be sourced), they can even lead to project termination and failure. Risks that must be considered relative to student involvement include estimation error, cost creep, damage and rework costs, and costs associated with meeting schedule requirements.

7.3.4.1 Cost Estimation Error

Cost estimation error closely mirrors schedule estimation error. It occurs when the level of cost required to be incurred for a given activity is different from the level forecast. While variation is expected, proper estimation should result in some tasks concluding with small overruns and others being completed under budget. Generally, an allowance for unexpected costs is included in the budget as a separate line item to allow the absorption of additional costs, should the project average out to a slight overrun. As with schedule estimation error, students who may be estimating costs for the first time (or may have limited domain experience, even if they have performed cost estimating before) may be prone to underestimate, due to ignoring complexity or inadvertently omitting various types of costs or specific costs.

7.3.4.2 Cost Creep

Cost creep is the cost component of scope creep. Scope creep occurs when changes are accepted without commensurate changes in budget and schedule. Due to student inexperience and other factors, scope creep is likely on student projects. If scope creep occurs, it is likely that cost creep will occur.

7.3.4.3 Damage and Rework

Damage and rework costs are incurred when hardware, facilities, supplies, or the item being created is damaged due to carelessness, accident, misuse, or otherwise. Damage and rework costs are likely on a student-involved project. First, the lack of

7.3 The Student Qualitative Undertaking Involvement Risk Model

a production environment designed for the repetitive production of an item means that construction and integration jigs will be set up on the fly. This may result in inadvertent loss of control, dropping, or the application of unwanted torques or pressure to parts or assemblies. Second, the lack of a repetitive production environment means that there is not a set of well-tested task instructions that can be followed. Third, supply and equipment limitations may result in jury rigging of various jig elements, making damage more likely. Fourth, horseplay or carelessness may result in damage. All of the aforementioned are exacerbated by having young and/or inexperienced individuals working on the project.

7.3.4.4 Buying Time

Costs can be incurred to resolve schedule issues. For example, a component could be purchased, at additional expense, to return the project to schedule or an external consultant could be hired to expedite a process. Due to this, schedule issues can become cost issues. Given how student involvement can exacerbate schedule risk, it would seem that student involvement would heighten the possibility of transferring schedule overruns to cost in order to hit a key deadline.

7.3.5 Risks Posed by Student Worker Involvement

Several risk factors are impacted so dramatically by student involvement as to deserve separate consideration from their standard counterparts. Each is now discussed in detail.

7.3.5.1 Scheduled Turnover

Scheduled turnover has a dramatic impact, but can be planned for. It is attributable to the fact that students only participate in a given effort for a period of time. When this participation ends the student may be unavailable to provide documentation or assistance related to his/her work on the project. As students become task-experts, if documentation is not stressed, understanding can be lost—or a key component of an integrated system can become unserviceable. Compounding this issue is the fact that many students are not adept in documenting their work and lack an understanding of the need for documentation and what needs to be documented. Mitigation strategies for this risk include knowledge distribution, stressing documentation throughout a project's lifecycle, and validating the usefulness of documentation, by requiring its use prior to a student worker's departure.

7.3.5.2 Unscheduled Turnover

Unscheduled turnover is a risk factor present in all types of organizations. As in corporate work, environments, medical, personal, and other factors may necessitate a worker's immediate departure from the workplace. Mitigation techniques for this class of risk include duplication (or responsibility distribution) of key roles, wide knowledge distribution, and stressing documentation and documentation validation.

7.3.5.3 Miss-commitment

Students' miss-commitment can be more problematic than the occurrence of turnover. With turnover, the project leader has knowledge of the current status of the team member. With miss-commitment, the individual is still present and ostensibly working on his/her assigned tasks; however, due to conflicting demands for limited time resources (and the academic trumping of most project duties) the student worker may not have time to make the requisite level of project progress. This is compounded by the cramming-centric work styles learned by many students, which lead to the belief that everything can be 'made up' at the last moment. With student miss-commitment, project leaders may not become aware of the issue, until investigating the cause of a key deadline being missed. Mitigation techniques for this class of risks include defining tasks to have demonstrable milestones, creating an environment where challenges are reported instead of obfuscated, and involving multiple individuals in key tasks.

7.3.5.4 Inexperience

Inexperience is, of course, a problem that is faced by numerous projects in every sphere. A team member may be new to the workforce or may lack experience in the specific areas required by a project. However, inexperience is a particular issue in student-centric projects as many students lack practical experience. This translates into misestimating and a lack of experience in problem resolution techniques. This class of risks can be mitigated by training students in the desired behaviors (e.g., how to estimate in a given sphere, how to deal with problems, etc.). This mitigation not only benefits the project but also prepares the students for workplace entry.

7.4 Extending the Model with Root Cause Analysis Techniques

The original SQUIRM model, presented in prior work [35], expanded upon the causal factors for standard risks, which could be exacerbated by student/inexperienced workers' involvement. While some discussion of the causality of the student worker-specific risks was included, these were not incorporated into the formal model. The SQUIRM-Extended Model (SQUIRM-E) adds these causal factors to

7.4 Extending the Model with Root Cause Analysis Techniques

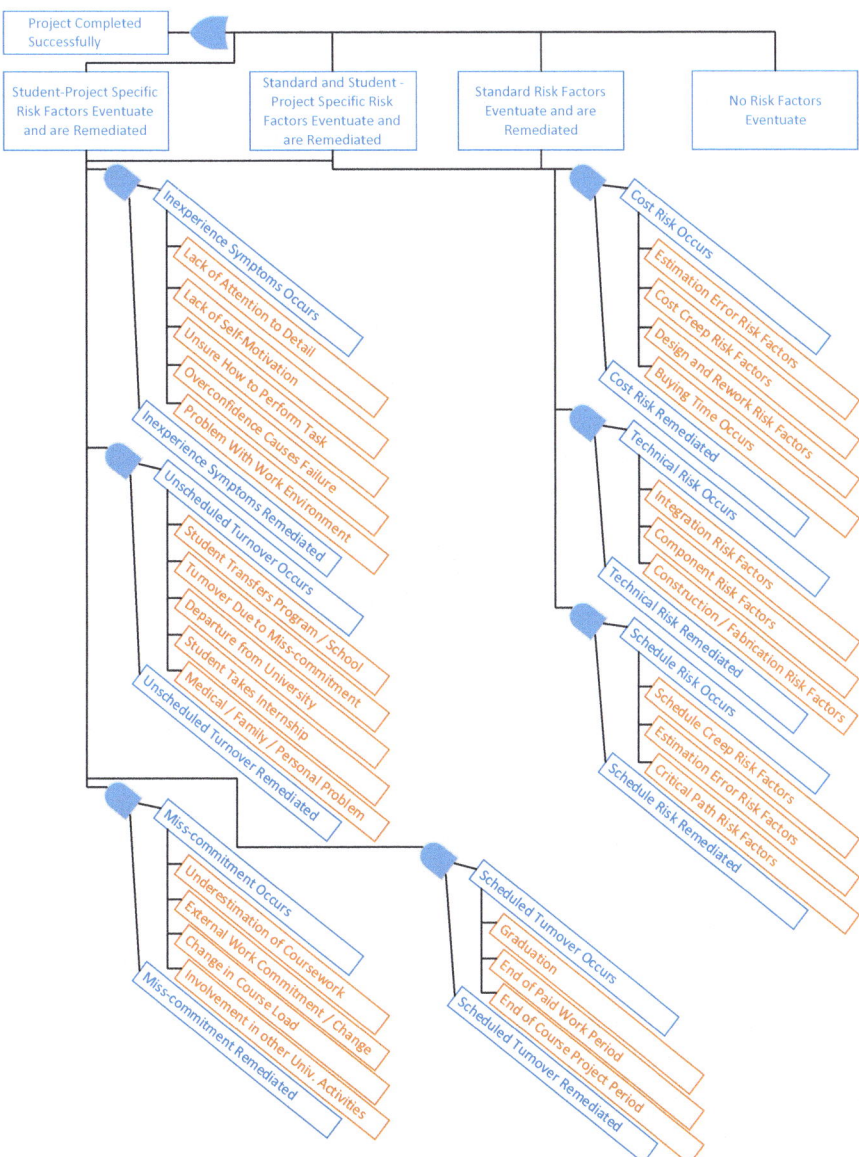

Fig. 7.2 SQUIRM-E model diagram

the model, as shown in Fig. 7.2. This addition is necessary to begin quantitative assessment using the model (which is discussed in a subsequent section). This section begins with a discussion of the value of the use of root cause analysis and then the rest of this section discusses the new elements of the SQUIRM-E model and expands upon the types of risks posed by them and their causes.

7.4.1 Root Cause Analysis

The premise of root cause analysis (RCA) is that a better understanding of the underlying factors of an exceptional occurrence (either positive or negative) facilitates a better understanding of how negative occurrences can be avoided in the future and positive occurrences brought about. Significant prior work exists in this area; a high level overview is provided by Rooney and Heuvel [62]. RCA has been used for process analysis [63], investigating medical error [64], and improving patient safety [65], as well as in analyzing and improving industrial safety and performance [66]. A discussion of several tools for RCA was presented by Doggett [67].

In the context of this work, RCA was used to assess why student-involved projects and student workers could have higher levels of risk actualization than a similar project not incorporating inexperienced workers. In prior work [35], this was applied to seek out causes that were specific to student (and other inexperienced) workers. In this chapter, RCA is used to decompose standard risk factors to assess the prospective contribution of inexperience and related factors on these risk areas.

RCA is not the only technique that could be used to assess these types of risks. However, it has several benefits. Unlike some other approaches, for example, it uses a bottom-up approach which makes it suitable for projecting risks instead of analyzing actualized risks. This is particularly valuable in the context of nonoperation risk analysis, where prior occurrences in a recurring process cannot be analyzed to project future risk factors and their likelihood. With RCA, the individual factors contributing to each type of prospective risk have been identified. These can, then, inform planning (in order to facilitate avoidance and mitigation) as well as be used to arrive at an understanding of the risk level of a project and its areas of particular risk. To perform RCA, prospective sources of the higher level risk factors previously presented were identified. These are described in greater detail throughout the remainder of this section.

7.4.2 Inexperience Symptoms Occur

The risk categories related to inexperience are a lack of attention to detail, lack of self-motivation, uncertainty as to how to perform a task, overconfidence that causes failure, and problems with the work environment. These are now discussed.

7.4.2.1 Lack of Attention to Detail

Student workers may lack an understanding of the importance of particular details of a task, lack an understanding of the actual details (i.e., what is a correct implementation at a detailed level versus an incorrect one), or may simply fail to pay the level of attention required. This may be exacerbated due to other time commitments

7.4 Extending the Model with Root Cause Analysis Techniques   113

(reducing the amount of time that can be devoted to these details and task performance), the level of strain that the student is under (particularly if the student lacks coping mechanisms), and other factors (such as the amount of time available during the semester).

7.4.2.2 Lack of Self-Motivation

Students (particularly lower level undergraduates) may not yet have developed the skills, habits, and work ethic required to self-motivate work when tasks seem unexciting or are in support of a longer-term goal. This may translate into unsatisfactory performance in terms of meeting deadlines, unsatisfactory work product, or other deficiencies. It may also trigger or contribute to other risk factors (such as miss-commitment if work piles up due to not starting things until there is an imminent due date).

7.4.2.3 Unsure of How to Perform Task

Students may be unsure of how to perform particular tasks or elements of a task. This may translate into delays waiting for clarification, attempts that result in wasted materials and time, obviously defective products or products with latent defects that may impair progress during later phases (e.g., integration, testing). This lack of understanding may decrease motivation, increase frustration, and delays may trigger other issues such as miss-commitment.

7.4.2.4 Overconfidence Causes Failure

Students may underestimate the difficulty of a task or overestimate their own capabilities. This can have several different symptoms, depending on when it occurs. First, it can cause issues with scheduling and costing. Students may underestimate the amount of time that will be required for learning how to perform a task, experimenting to gain understanding, and/or correcting less-than-acceptable products. They may also underestimate the amount of waste material that may be consumed by reattempts to fix defects.

Second, it can result in unsatisfactory performance in terms of meeting deadlines, unsatisfactory work product, or other deficiencies due to the aforementioned scheduling and the reality of performance conflicting, or a lack of understanding of what an acceptable product is, triggering a need for significant rework. This may translate into delays waiting for clarification, attempts that result in wasted materials and time, obviously defective products, or products with latent defects which may impair progress during later phases (e.g., integration, testing). These issues may trigger other risk factors such as miss-commitment, decreased motivation, and increased frustration.

Third, this may result in students responding negatively to feedback, as they think that it is unnecessarily critical (based on their inaccurate assumptions about their own capabilities and what constitutes an acceptable level of performance). This may also increase frustration, decrease motivation, and potentially trigger other issues, such as turnover.

7.4.2.5 Problem with Work Environment

Student workers may lack an understanding of how to cope with difficulties in the workplace environment. For example, they may not understand how to deal with a poor manager (and the, particularly if a student, manager may lack the skills and understanding required to resolve this conflict). They may also lack the skills required to resolve workplace conflict or to collaborate with others in the work environment. This can potentially trigger miss-commitment, if work is left to pile up while issues are being resolved, or if unscheduled turnover occurs.

7.4.3 Unscheduled Turnover Occurs

Unscheduled turnover can be caused by a student transferring between degree programs and colleges/universities, as a result of miss-commitment, because of a student's departure from the university, or even by a student taking an internship or a medical, family, or other personal problem. Each is now discussed.

7.4.3.1 Student Transfers Program/School

In the context of their educational pursuits, students make decisions in light of what they perceive as their own best interests (which may consider short- and/or long-term goals). The inflexibility of the semester system may limit students' ability to provide notice (even for a paid position), should they decide to transfer between schools or programs. They may also lose interest at the point that they realize that program participation is no longer supporting their goals (framed now in terms of their new school/department). This may result in low or no-notice turnover.

7.4.3.2 Turnover due to Miss-commitment

Students may miss-commit (reasons for this are discussed subsequently). If this miss-commitment becomes an acute problem, students may terminate their involvement in paid and/or unpaid extracurricular activities in deference to their immediate academic time needs. This may occur with low or no notice or it may simply result in the student failing to show up (without any sort of explanation).

7.4 Extending the Model with Root Cause Analysis Techniques   115

7.4.3.3 Departure from University

Students may leave (or be dismissed from) the university for a wide variety of reasons. This may also result in low or no-notice turnover.

7.4.3.4 Student Takes Internship

Students may decide to pursue an internship to increase their skills and/or postgraduation employment opportunities. Internships may pay more than on-campus employment and generally offer work experience benefits and prospective employer contact that on-campus employment cannot. Students may begin an internship with little or no notice (as employers may offer internships at the last minute to meet their needs and funding capabilities); in many cases, however, internships can be a planned absence and a student may be able to/decide to return to the project after its completion.

7.4.3.5 Medical/Family/Personal Problem

Like any worker, students may suffer from medical family or other personal problems. These may be intensified by students' lack of coping skills and/or the lack of a need to maintain an income, even in the face of a major medical condition. Notice levels, the potential for students to return to the project upon the resolution of the issue and the duration of the issue will, obviously, vary significantly based on the nature of the issue.

7.4.4 Scheduled Turnover Occurs

Scheduled turnover is an expected occurrence at a college or university. It can be caused by student graduation, the end of a paid (e.g., extramurally funded) work period, or the end of a course project period. Each is now considered.

7.4.4.1 Graduation

Students enroll in a university with their departure planned (unlike a typical work environment where employees may not plan to make a career out of a job, but also look at it as something to pursue for an indeterminate period of time). Graduation, fortunately, will be an occurrence that is known well in advance and can be planned for to ensure proper handover. Students, however, may fail to notify project leaders (either due to a presumption that they should be notified by some other means or to avoid less-interesting handover activities) and/or have a declining level of interest (particularly after they have secured a job or admission into another program for graduate studies, etc.) that may reduce the ability to conduct and/or the quality of handover activities.

7.4.4.2 End of Paid Work Period

Research grant (or other funding source) work may have a definite cutoff point after which no additional funding is available to continue a position. This creates a known date-of-departure for a student from a project (or a transition from a paid role to continuation on a volunteer basis). This should be known to the investigator (and thus not suffer from the aforementioned failure-to-notify problem) and be able to be planned for. Students may lose interest and/or change their final day if they find an alternate position, as they approach their known final days.

7.4.4.3 End of Course Project Period

Course projects, like paid work periods, have definite (and known-to-the investigator) end dates. A desire to receive a good final grade, however, may keep students motivated until the end of the period.

7.4.5 Miss-commitment

Miss-commitment is to be expected with students who may be unable to gauge the level of work required both from their academic, paid work and extramural pursuits. Miss-commitment, thus, can occur due to students' underestimation of coursework time commitments, an external work commitment commencing or changing, a change in a student's course load, and/or involvement in other university activities. These are now considered.

7.4.5.1 Underestimation of Coursework

Students may overcommit to extramural projects or paid on-campus project work, based on an underestimation of the level of time required for their coursework. This may result in delays, turnover, or impaired quality.

7.4.5.2 External Work Commitment/Change

Students who are working on a project in either a paid or volunteer basis may have jobs outside the project or may seek/take a job based on the benefits it may provide (e.g., work experience, employer contact) or due to their personal financial situation. This may result in low or no-notice changes in project involvement levels, turnover, or a decline in product quality.

7.4.5.3 Change in Course Load

Students may change the number or selection of courses they are taking during the semester and this may change somewhat from semester to semester. This may result in turnover, delays, or quality impairment.

7.4.5.4 Involvement in Other University Activities

Students may decide to pursue other university extracurricular activities in addition to or instead of the project, or the level of involvement required for (or desired in) these activities may change, reducing the students' level of involvement in the project and/or causing delays, quality problems, or turnover.

7.5 Differences Between and Choosing Between Using SQUIRM and SQUIRM-E

With both the SQUIRM framework and its extension presented, the two can now be compared. This section reviews the differences between SQUIRM and its extension, SQUIRM-E. It discusses the benefits of using one versus the other across multiple scenarios.

7.5.1 Discussion of the Differences Between SQUIRM and SQUIRM-E

The fundamental difference between SQUIRM and SQUIRM-E is the addition, in SQUIRM-E, of the decomposition of standard risk classes in order to also consider risk sources attributable to student and inexperienced workers. This has resulted in two models, each of which is better suited for certain applications (as compared to the other). The remainder of this section considers specific benefits of using one model over the other. It begins by discussing the comparative simplicity presented by SQUIRM, versus SQUIRM-E, and where this simplicity may be valuable. Next, it discusses how SQUIRM-E leans further towards student workers, making SQUIRM more suitable for use or adaptation to nonstudent, inexperienced workers (or students in contexts where the student status is less relevant). Finally, logistical considerations such as project size and assessor environment familiarity are discussed before a concluding discussion regarding model selection.

7.5.2 Comparative Simplicity

The SQUIRM framework, by abstracting the root causes of the student-specific risk types into larger categories, is comparatively easier to work with. This is particularly useful in cases where real numbers for these risk types are unknown and cannot be accurately estimated or where data has been collected without sufficient granularity for use with the more granular model. Alternately, those estimating without data may prefer the more detailed model, as it allows them to consider the risk, likelihood, and impact for specific prospective problems, without having to consider whole categories at one. The use of the SQUIRM-E framework, thus, would correspond to a bottom-up risk identification strategy, while the SQUIRM framework (for student-specific risk types) would correspond to a top-down risk identification and assessment approach.

7.5.3 Types of Inexperienced Workers

While the SQUIRM model contains elements that may be useful for all areas of inexperience, the elaborations in SQUIRM-E have been targeted specifically at student workers (with a particular focus towards student workers working in the context of a university environment). The further that the actual situation diverges from this, the less valuable the SQUIRM-E elaborations may be. Alternately, one might use these as a starting point, removing (and/or replacing) irrelevant topics and making changes as needed to relevant ones that have an incorrect focus for the scenario under consideration.

7.5.4 Project Size

For smaller projects or projects that are less critical, there may be less need for and resources with which to perform risk management. In these cases, the use of the simpler model (and, in fact, even simplifying the SQUIRM framework to remove the third-level error sources) may be prudent.

7.5.5 Familiarity with Particulars of Student Work Environment

Those with greater familiarity with the risks and nature of the student-involved work environment may find less need for the additional granularity of the SQUIRM-E model. However, as some risk types occur infrequently, heuristic models based on past experiences may oversimplify actual risk levels. Alternately, nonuniversity employers who are less familiar with the particulars of student worker risks may desire to use a modified version of the full SQUIRM-E model. This adaptation is discussed in a subsequent section.

7.5.6 Choosing a Model

While the two models are not that dissimilar, the selection of a model should be based on the complexity of the project as well as particular needs related to assessing student-status-attributable risk factors. Choosing the incorrect version of the model to use may result in oversimplification, under or overstatement of risks, and/or unnecessary work.

7.6 Application

If similar projects were planned in the future, they could use SQUIRM or SQUIRM-E as appropriate (see above), following a five-step approach.

First, the nature of the project must be defined. A discussion of this is beyond the scope of this chapter; however, several common frameworks exist, including those by Wertz et al. [68] and Fortescue et al. [69]. A simplified version for small high altitude ballooning projects (which could be adapted to other aerospace projects) has also been proposed [70]. These frameworks incorporate risk analysis in different ways; however, this process—using SQUIRM/SQUIRM-E—should involve the following four steps.

Second, areas of student (inexperienced staff) involvement, areas impacted by student involvement, and areas not impacted by or involving students should be identified. The use of the SQUIRM/SQUIRM-E model is appropriate for the first two areas; the last one should use conventional risk assessment and management techniques.

Third, a granularity level of risk assessment must be determined, based on the scale and nature of the project. Risk could be assessed at the whole project level or at any logical division level thereunder. The granularity level need not be consistent; thus, areas of higher risk or risk impact could be assessed at higher levels of granularity than less risky or impactful areas.

Fourth, for each unit of assessment, risk factors should be identified. This will involve application/task-specific brainstorming as well as reviewing the student/inexperienced worker-attributable factors presented by the SQUIRM model. For each factor, a likelihood and impact level should be estimated (based on historic data, experience, or other technique).

Finally, any summative assessment should be performed. This may include combining risk data from subtasks into task level assessments (or from tasks into project level assessment), evaluating student/inexperienced worker participation value and comparing project level assessments.

The foregoing can be performed qualitatively or quantitatively. Quantitative analysis is discussed in greater detail in the subsequent section.

7.7 Quantifying the Model

While the discussion up to this point has been qualitative, both the SQUIRM and SQUIRM-E models lend themselves to being used with quantitative data, if it is available. Figure 7.3 demonstrates how the identified risk areas, along with mitigation/response strategies identified using the SQUIRM/SQUIRM-E model, can be used to assess the weighted (by likelihood of occurrence) risk impact levels for particular risk sources and for the project overall. The overall project risk levels may serve to facilitate comparison between projects (in conjunction with other metrics such as project importance and cost).

7.7.1 Risk Assessment

Risks are assessed in terms of both their likelihood of occurrence and the magnitude of impact that they may have if they eventuate. Risks may be assessed based on probabilities, if sufficient historical data exists or a probabilistic model is known or

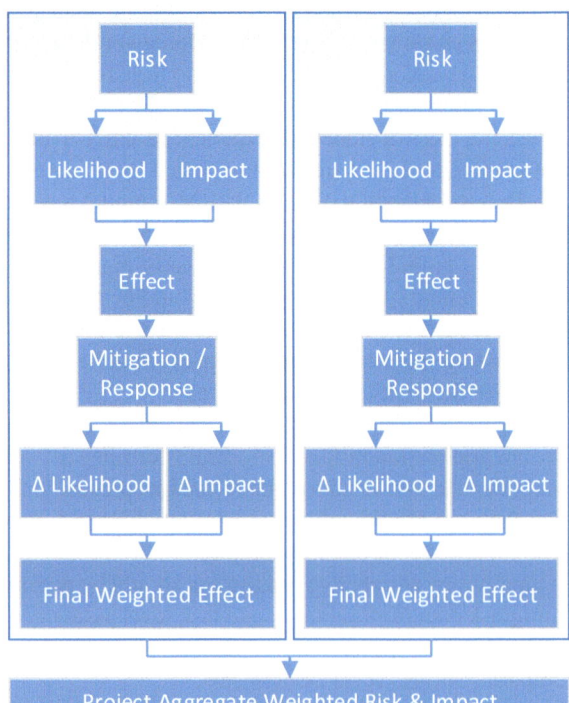

Fig. 7.3 Quantifying SQUIRM and SQUIRM-E

can be inferred, or they can be categorized (with approximate average probabilities, used to facilitate quantitative comparisons).

The impact can, similarly, be quantified in terms of time, resource, and cost (which may be combined into a single cost metric), if data is available. Alternately, they can be categorized and an average value used.

7.7.2 Mitigation/Response Assessment

The risk effect may be altered by the existence (or development) of mitigation and response strategies. Mitigation strategies may reduce likelihood, impact, or both, while response strategies focus solely on reducing impact. The change created by the existence of one or more of these strategies should be considered. Again, actual numbers or classifications and average values can be used for this assessment.

7.7.3 Combining for Result

The risk effect and mitigation/response change are combined for each risk factor. Then (if multiple risk factors are present), the final weighted effects are combined, to produce an aggregated risk impact value for the project. It is important, when using this approach, that all values use a comparable scale (e.g., combining average and historical cost values should be done carefully to avoid over- or understatement of risks). If risk values are being used to compare projects, then the need for a common scale extends to all items being compared. Thus, it is ideal (but often not practical) to use historical data and (inflation and other factor adjusted) real costs, as this facilitates direct comparison.

7.7.4 Data for Model Parameters

One particular challenge in the use of SQUIRM or SQUIRM-E quantitatively is the collection of the parameters which are required in order to perform the quantitative analysis. Problematically, this data likely varies on an application-specific basis (or general data would need to be validated for application-specific use). While, for small satellites, some relevant data has been collected by Brumbaugh and Lightsey [71], and they are collecting data [72] to facilitate a more robust analysis, this doesn't cover all areas required by this model nor does it help those attempting to assess risk in other application areas. For areas and applications where this data is not available, it will need to be estimated based on past experience and other available information. The collection of data specific to particular applications is an area for future work.

7.8 Value Model for Inexperienced Workers

The foregoing may lead one to question the value of using inexperienced workers (particularly students) on any project of particular importance. Would the students/junior employees not be better served (and better serve others) by gaining experience through nonimpactful learning exercises instead of work on real projects (which could be negatively impacted)? This section considers the value of student (and other inexperienced) workers. Figure 7.4 presents a diagram of the considerations.

7.8.1 Cost of Inexperienced/Student Workers

The cost of inexperienced and student workers is aptly identified by the SQUIRM and (to a greater extent) SQUIRM-E models. Clearly, each prospective risk may impair a project (if it eventuates) incurring time, productivity impairment (including productivity impairment of other more senior workers that may need to help rectify student/inexperienced worker mistakes), material, and goodwill costs. Somewhat (in many cases) offsetting, this is the lower wage levels paid to student/inexperienced staff. Thus, for tasks that these individuals can learn to perform effectively and with minimal (or comparable to more experienced staff) oversight, a cost saving may be enjoyed. The assignment of junior staff to these types of tasks, however, may impair their learning process and prevent them from gaining (or decrease the speed of them gaining) skills that could make them more valuable to their current and future prospective employers.

7.8.2 Training Benefits

The proverbial adage of "killing two birds with one stone" can be used in an attempt to justify the use of student/unskilled workers on real projects. If students/unskilled workers can be productively contributing to a project while also gaining experience, it would seem that two types of benefits are being gained for a single cost. While

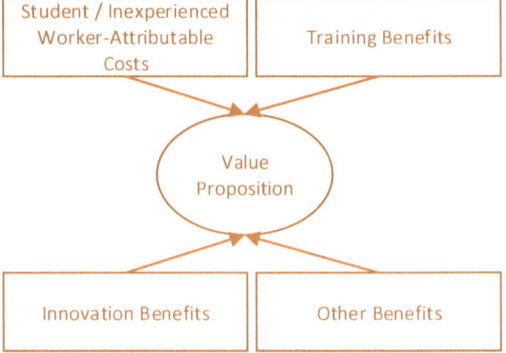

Fig. 7.4 Value proposition for the use of inexperienced and student workers

this may certainly be true in some (perhaps many) cases, the oversimplification of the cost model (i.e., the consideration of a "single" cost) may be inaccurate. Costs may be higher to facilitate the student/inexperienced worker participation, which should be taken into account in the comparison.

7.8.3 Discontinuous Innovation Benefits

One area where student/inexperienced workers may offer particular benefit is in identifying sources of discontinuous innovation. These workers, who may not fully understand where the proverbial "box" is, may be well suited to think outside of it. Swartwout [38, 39] identifies this, for example, as a key benefit of "university-class" small spacecraft programs: the higher level of risk tolerance and the presence of the junior staff make these types of missions well suited to trying innovative ideas and identifying areas for innovation in operations.

7.9 Discussion of the Differences Between Student Volunteers, Paid Student Workers, Interns, and Junior Employees

It has been stated, previously, that the SQUIRM and SQUIRM-E frameworks can be used to address risks across several different types of junior employees; however, the risk factor impacts posed by these different groups are dissimilar. This section begins the process of considering the differences between the multiple types of workers that the SQUIRM/SQUIRM-E models could be applied to (in some cases with limited modifications). The particulars of each worker type are now discussed; this includes student volunteers, paid student workers, interns, and junior employees.

7.9.1 Student Volunteers

Student volunteers will (correctly) view their participation as at will. If they are interested, see benefits being provided and have time, they will continue working on the project. If they lose interest, feel that they are not receiving (or have already received all applicable) benefits, or are confronted with other draws on their time, they will stop. Retention of students from semester to semester may be difficult, as they may perceive participation as an opt-in activity (like joining a club or taking a class), where a participation decision is made anew each semester. They may fail to realize or understand the impact of their change in participation status on others who have also donated their time to provide benefit to them or the cost of the time committed to their training by paid staff, etc.

7.9.2 Paid Student Workers

Paid student workers may be more committed, as they are receiving another source of benefit (pay) over and above what is received by volunteers. However, in the context of the comparatively large amounts of money that they are paying (or which is being paid on their behalf) to attend school, they may see little difference between the paid and unpaid positions in terms of any sense of commitment or longer-term responsibility. Pay, thus, may overcome (or assist in rectifying) lack of interest issues, but may not assist with semester-to-semester turnover issues or commitment in the face of other time draws.

7.9.3 Interns

Interns (in the case of nonuniversity employers) may see a multifaceted benefit which may cause particular (comparative) commitment. The intern may be earning credit for his/her participation, getting paid, gaining experience, and gaining an opportunity to demonstrate his/her capabilities to a prospective employer. The foregoing (particularly if the intern sees the employer as a desirable place to seek postgraduation employment) may cause interns to place the internship among their highest priorities, overcoming most of the common (controllable) risk factors and creating a particularly high level of diligence. Interns may or may not have ongoing coursework during the internship period (the lack thereof reducing another set of risk factors). As a generally fixed-term period of employment, however, scheduled turnover is expected.

7.9.4 Junior Employees

Junior employees may see performance as critical to their future livelihood; however, this perception may not always be the case (even if it is accurate, it may not be perceived or employment may be perceived as an entitlement). While most will want to set their careers off on a 'good foot,' others may find the change in structure (more or less control, different control structures, and a need to be self-starting) problematic and not know how to function effectively under the changed structure. Employees may also be looking for new positions, if they take a position that is not of their liking simply to 'pay the bills' and may lack the professional discipline to continue to perform while in a job they dislike (or which they are not particularly excited about).

7.10 Conclusions and Future Work

This chapter has presented SQUIRM-E a version of the SQUIRM framework that adds additional assessment criteria related to student-specific risk types. It has presented an analytical framework for assessing risk factors, relevant to student and inexperienced workers quantitatively, and evaluating the value of the use of a student/inexperienced worker on a given project. A limited extrapolation to nonstudent workers has been discussed.

Future work will involve the enhancement of the quantitative models presented as well as the collection of a data set to begin to characterize these common risk areas for various classes of projects. It will also involve the development of a SQUIRM-E-based model for junior employees that replaces student-specific factors with those more appropriate to junior employees.

References

1. Straub, J. 2014. Extending the student qualitative undertaking involvement risk model. *Journal of Aerospace Technology and Management* 6(3): 333–352.
2. Snell, K.D. 1996. The apprenticeship system in British history: The fragmentation of a cultural institution. *History of Education* 25(4): 303–321.
3. Elbaum, B. 1989. Why apprenticeship persisted in Britain but not in the United States. *Journal of Economic History* 49(2): 337–349.
4. Straub, J., J. Berk, A. Nervold, and D. Whalen. 2013. OpenOrbiter: An interdisciplinary, student run space program. *Advances in Education* 2: 4–10.
5. Mountrakis, G., and D. Triantakonstantis. 2012. Inquiry-based learning in remote sensing: A space balloon educational experiment. *Journal of Geography in Higher Education* 36(3): 385–401.
6. Nordlie, J., and R. Fevig. 2011. Blending research and teaching through high-altitude balloon projects. In *Proceedings of the 2nd Annual Academic High Altitude Conference*. Ames, IA.
7. Hall, S.R., I. Waitz, D.R. Brodeur, D.H. Soderholm, and R. Nasr. 2002. Adoption of active learning in a lecture-based engineering class. Presented at Proceedings of the 32nd Annual Frontiers in Education Conference.
8. Brodeur, D.R., P.W. Young, and K.B. Blair. 2002. Problem-based learning in aerospace engineering education. Presented at Proceedings of the 2002 American Society for Engineering Education Annual Conference and Exposition.
9. Mathers, N., A. Goktogen, J. Rankin, and M. Anderson. 2012. Robotic mission to mars: Hands-on, minds-on, web-based learning. *Acta Astronautica* 80: 124–131.
10. Pollard, C.E. 2012. Lessons learned from client projects in an undergraduate project management course. *Journal of Information Systems Education* 23(3): 271–282.
11. Dahlgren, M.A., and L. Dahlgren. 2002. Portraits of PBL: Students' experiences of the characteristics of problem-based learning in physiotherapy, computer engineering and psychology. *Instructional Science* 30(2): 111–127.
12. Duch, B.J. 1996. Problem-based learning in physics: The power of students teaching students. *Journal of College Science Teaching* 15(5): 326–329.
13. Correll, N., R. Wing, and D. Coleman. 2013. A one-year introductory robotics curriculum for computer science upperclassmen. *IEEE Transactions on Education* 56(1): 54–60. doi:10.1109/TE.2012.2220774.
14. Broman, D., K. Sandahl, and M. Abu Baker. 2012. The company approach to software engineering project courses. *IEEE Transactions on Education* 55(4): 445–452. doi:10.1109/TE.2012.2187208.
15. Roh, K.H. 2003. Problem-based learning in mathematics. *ERIC Clearing House for Science Mathematics and Environmental Education* 2004–2003.
16. Okudan, G.E., and S.E. Rzasa. 2006. A project-based approach to entrepreneurial leadership education. *Technovation* 26(2): 195–210.

17. Saunders-Smits, G.N., P. Roling, V. Brügemann, N. Timmer, and J. Melkert. 2012. Using the engineering design cycle to develop integrated project based learning in aerospace engineering. Presented at International Conference on Innovation, Practice and Research in Engineering Education.
18. Jayaram, S., L. Boyer, J. George, K. Ravindra, and K. Mitchell. 2010. Project-based introduction to aerospace engineering course: A model rocket. *Acta Astronautica* 66(9): 1525–1533.
19. Qidwai, U. 2011. Fun to learn: Project-based learning in robotics for computer engineers. *ACM Inroads* 2(1): 42–45.
20. Bütün, E. 2005. Teaching genetic algorithms in electrical engineering education: A problem-based learning approach. *International Journal of Electrical Engineering Education* 42(3): 223–233.
21. de-Camargo-Ribeiro, L.R. 2008. Electrical engineering students evaluate problem-based learning (PBL). *International Journal of Electrical Engineering Education* 45(2): 152–161.
22. Robson, N., I.S. Dalmis, and V. Trenev. 2012. Discovery learning in mechanical engineering design: Case-based learning or learning by exploring? Presented at 2012 ASEE Annual Conference.
23. Coller, B.D., and M.J. Scott. 2009. Effectiveness of using a video game to teach a course in mechanical engineering. *Computers & Education* 53(3): 900–912.
24. Hayne, S., M. Pendergast, and C. Smith. 2012. Team composition, knowledge and collaboration. Presented at SAIS 2012 Proceedings.
25. Sulaiman, S.A., H.M. Git, and A.M.A. Rani. 2010. Engineering team project course for development of teamwork skills. Presented at Proceedings of the 3rd Regional Conference on Engineering Education and Research in Higher Education.
26. Jackson, D., and P. Hancock. 2010. Non-technical skills in undergraduate degrees in business: Development and transfer. *Education Research and Perspectives* 37(1): 52–84.
27. Doppelt, Y. 2003. Implementation and assessment of project-based learning in a flexible environment. *International Journal of Technology and Design Education* 13(3): 255–272.
28. Ayob, A., R.A. Majid, A. Hussain, and M.M. Mustaffa. 2012. Creativity enhancement through experiential learning. *Advances in Natural and Applied Science* 6(2): 94–99.
29. Bauerle, T.L., and T.D. Park. 2012. Experiential learning enhances student knowledge retention in the plant sciences. *HortTechnology* 22(5): 715–718.
30. Simons, L., L. Fehr, N. Blank, H. Connell, D. Georganas, D. Fernandez, and V. Peterson. 2012. Lessons learned from experiential learning: What do students learn from a practicum/internship? *International Journal of Teaching and Learning in Higher Education* 24(3): 325–334.
31. Nagda, B.A., S.R. Gregerman, J. Jonides, W. von Hippel, and J.S. Lerner. 1998. Undergraduate student-faculty research partnerships affect student retention. *The Review of Higher Education* 22(1): 55–72.
32. Hotaling, N., B.B. Fasse, L.F. Bost, C.D. Hermann, and C.R. Forest. 2012. A quantitative analysis of the effects of a multidisciplinary engineering capstone design course. *Journal of Engineering Education* 101(4): 630–656.
33. Fasse, B., N. Hotaling, C. Forest, L. Bost, and C. Hermann. 2012. The case for multi-disciplinary capstone design: A quantitative analysis of impact on job placement and product quality. Presented at Proceedings of the Biomedical Engineering Society (BMES) 2012 Annual Meeting. http://pbl.gatech.edu/wp-content/uploads/2013/05/fasse_bmes_2012.pdf.
34. Gilmore, M. 2013. Improvement of STEM education: Experiential learning is the key. *Modern Chemistry & Applications* 1: e109.
35. Straub, J., R. Fevig, J. Casler, and O. Yadav. 2013. Risk analysis & management in student-centered spacecraft development projects. In *Proceedings of the 2013 Reliability and Maintainability Symposium*. Orlando, FL.
36. Straub, J., and D. Whalen. 2013. An assessment of educational benefits from the OpenOrbiter space program. *Education Sciences* 3(3): 259–278.
37. Straub, J., D. Whalen, and R. Marsh. 2014. Assessing the value of the OpenOrbiter program's research experience for undergraduates. *Sage Open* 2014.

38. Swartwout, M. 2011. AC 2011-1151: Significance of student-built spacecraft design programs it's impact on spacecraft engineering education over the last ten years. Presented at Proceedings of the American Society for Engineering Education Annual Conference. http://www.asee.org/file_server/papers/attachment/file/0001/1307/paper-final.pdf.
39. ———. 2004. University-class satellites: From marginal utility to 'disruptive' research platforms. Presented at Proceedings of the 18th Annual AIAA/USU Conference on Small Satellites.
40. Larsen, J.A., and J.D. Nielsen. 2011. Development of CubeSats in an educational context. Presented at 2011 5th International Conference on Recent Advances in Space Technologies (RAST).
41. Larsen, J.A., J.F.D. Nielsen, and C. Zhou. 2013. Motivating students to develop satellites in problem and project-based learning (PBL) environment. *International Journal of Engineering Pedagogy* 3(3): 11–17.
42. Straub, J. 2013. OpenOrbiter: Analysis of a student-run space program. Presented at Proceedings of the 64th International Astronautical Congress.
43. Zydney, A.L., J.S. Bennett, A. Shahid, and K.W. Bauer. 2002. Impact of undergraduate research experience in engineering. *Journal of Engineering Education* 91(2): 151–157.
44. ———. 2002. Faculty perspectives regarding the undergraduate research experience in science and engineering. *Journal of Engineering Education* 91(3): 291–297.
45. Prince, M.J., R.M. Felder, and R. Brent. 2007. Does faculty research improve undergraduate teaching? An analysis of existing and potential synergies. *Journal of Engineering Education* 96(4): 283–294.
46. Botterill, L., and N. Mazur. 2004. Risk and risk perception: A literature review. *Project no. BRR-8A, Rural Industries Research and Development Corporation, Barton.*
47. Slovic, P., B. Fischhoff, and S. Lichtenstein. 1982. Why study risk perception? *Risk Analysis* 2(2): 83–93.
48. Boholm, A. 1998. Comparative studies of risk perception: A review of twenty years of research. *Journal of Risk Research* 1(2): 135–163.
49. Mitchell, V. 1995. Organizational risk perception and reduction: A literature review. *British Journal of Management* 6(2): 115–133.
50. Weinstein, N.D. 1984. Why it won't happen to me: Perceptions of risk factors and susceptibility. *Health Psychology* 3(5): 431.
51. Deery, H.A. 2000. Hazard and risk perception among young novice drivers. *Journal of Safety Research* 30(4): 225–236.
52. Levinson, R.A., J. Jaccard, and L. Beamer. 1995. Older adolescents' engagement in casual sex: Impact of risk perception and psychosocial motivations. *Journal of Youth and Adolescence* 24(3): 349–364.
53. Cohn, L.D., S. Macfarlane, C. Yanez, and W.K. Imai. 1995. Risk-perception: Differences between adolescents and adults. *Health Psychology* 14(3): 217.
54. Steinberg, L. 2004. Risk taking in adolescence: What changes, and why? *Annals of the New York Academy of Sciences* 1021(1): 51–58.
55. DeJoy, D.M. 1992. An examination of gender differences in traffic accident risk perception. *Accident Analysis & Prevention* 24(3): 237–246.
56. Rippl, S. 2002. Cultural theory and risk perception: A proposal for a better measurement. *Journal of Risk Research* 5(2): 147–165.
57. Sjöberg, L. 2000. Factors in risk perception. *Risk Analysis* 20(1): 1–12.
58. Wildavsky, A., and K. Dake. 1990. Theories of risk perception: Who fears what and why? *Daedalus* 119(4): 41–60.
59. Ronan, K.R., and D.M. Johnston. 2001. Correlates of hazard education programs for youth. *Risk Analysis* 21(6): 1055–1064.
60. Weber, E.U., and R.A. Milliman. 1997. Perceived risk attitudes: Relating risk perception to risky choice. *Management Science* 43(2): 123–144.
61. Renn, O. 1998. The role of risk perception for risk management. *Reliability Engineering & System Safety* 59(1): 49–62.

62. Rooney, J.J., and L.N.V. Heuvel. 2004. Root cause analysis for beginners. *Quality Progress* 37(7): 45–56.
63. Weidl, G., A. Madsen, and S. Israelson. 2005. Applications of object-oriented Bayesian networks for condition monitoring, root cause analysis and decision support on operation of complex continuous processes. *Computers & Chemical Engineering* 29(9): 1996–2009.
64. Iedema, R.A.M., C. Jorm, D. Long, J. Braithwaite, J. Travaglia, and M. Westbrook. 2006. Turning the medical gaze in upon itself: Root cause analysis and the investigation of clinical error. *Social Science and Medicine* 62(7): 1605–1615.
65. Neily, J., G. Ogrinc, P. Mills, R. Williams, E. Stalhandske, J. Bagian, and W.B. Weeks. 2003. Using aggregate root cause analysis to improve patient safety. *Joint Commission Journal on Quality and Safety* 29(8): 434–439.
66. Carroll, J.S., J.W. Rudolph, and S. Hatakenaka. 2002. Lessons learned from non-medical industries: Root cause analysis as culture change at a chemical plant. *Quality & Safety in Health Care* 11(3): 266–269.
67. Doggett, A.M. 2004. A statistical comparison of three root cause analysis tools. *Journal of Industrial Technology* 20(2): 2–9.
68. Wertz, J.R., D.F. Everett, and J.J. Puschell. 2011. *Space Mission Engineering: The New SMAD*. Hawthorne, CA: Microcosm Press.
69. Fortescue, P., G. Swinerd, and J. Stark. 2011. *Spacecraft Systems Engineering*, 4th ed. West Sussex: Wiley.
70. Straub, J., and R. Fevig. 2012. Formalizing mission analysis and design techniques for high altitude ballooning. Presented at Proceedings of the 3rd Annual High Altitude Conference.
71. Brumbaugh, K.M., and E.G. Lightsey. 2013. Application of risk management to university CubeSat missions. *Journal of Small Satellites* 2(1): 147–160.
72. ———. 2014. CubeSat Mission Risk Survey—A call for response from the small satellite community. *Journal of Small Satellites* 2: 83–84.

Chapter 8
Setting Educational Goals and Formative Assessment

This chapter begins the discussion of the educational aspects of small spacecraft programs (which continues in Chaps. 9 and 10). In this chapter, educational goals and formative assessment are discussed. Chapter 9 focuses on summative assessment, and Chap. 10 discusses the results of prior work.

The subsequent sections begin by providing an overview of prospective educational benefits. Then, a discussion of the difference between an educational program or one with educational benefits is provided. Next, the process of setting educational goals is discussed. After this a discussion of how to determine students' reasons for (particularly unpaid) participation and the use of this data is presented. Finally, the process of determining whether program and student goals are being met and how to take corrective action is considered, before concluding.

8.1 Educational Benefits: Overview

There are numerous instances of active, project (also known as problem)-based learning (PBL) being utilized to teach STEM skills. Several of these are in the space or near-space domain and are now discussed. Mathers et al. [5], for example, discuss how problem-based learning and a simulated robotic Mars mission are used at the Victorian Space Science Education Centre (VSSEC) to teach analytical and critical thinking skills. VSSEC provides educational experiences for teachers, pre-service teachers, and primary and secondary students. Two exercises, one designed for students in the fifth and sixth grade and the other designed for ninth and tenth

---

This chapter is based on, revises, and extends the papers "OpenOrbiter: A Low-Cost, Educational Prototype CubeSat Mission Architecture" [1], "OpenOrbiter: An Interdisciplinary, Student Run-Space Program" [2], "Extending the Student Qualitative Undertaking Involvement Risk Model" [3], and "Student Expectations from Participating in a Small Spacecraft Development Program" [4]. Some content in this chapter has also been previously summarized in [56].

grade students, are constructed around a simulated control room, limited-capability rover, and a simulated Martian environment.

Mountrakis and Triantakonstantis [6] present an inquiry-based approach to remote sensing, in the near-space environment. This work, conducted as an educational exercise for an introductory remote sensing course at the State University of New York College of Environmental Science and Forestry in Syracuse, tasked students with the creation of a high altitude balloon-based remote sensing platform. This low-budget experiment (students were allocated only $300) demonstrated the utility of student exploration of a problem space and the identification of technical solutions. Mountrakis et al. attribute a near doubling of the performance gain, compared to the control group of nonparticipating students, to student participation in the inquiry-based, active learning exercise.

Brodeur et al. [7] discuss the incorporation of problem-based learning into the undergraduate aeronautics and astronautics curriculum at the Massachusetts Institute of Technology. They proffer that the incorporation of problem-based learning has been very successful. A four-level classification system for the incorporation of problem-based learning into courses is presented. It includes exercises ranging from problem sets (structured exercises with known solutions) to minilabs (structured, short laboratory exercises), to macrolabs (multiweek to full-term projects) to capstone labs (complex design challenges). They suggest the need for learning to become "an act of discovery" and to simulate real-world and/or research environments and proffer that successful implementation "is contingent upon the design of good problems." Quantitative survey-based data indicates that the hands-on experiences were generally seen as very effective, relevant, and worthwhile by students.

Haruyama et al. [8] discuss a collaborative international program run at Keio University in Japan. This program, entitled the Active Learning Project Sequence (ALPS), is a 6-month exercise where students develop engineering solutions that respond to a selected central theme. They propose that this approach combines systems engineering practices with "design thinking," a key combination for solving complex system style problems. The program is based on a "V-model" approach to engineering, where design progresses from general concept brainstorming to low level design and testing progresses from unit level testing through to system testing. The "V-model" is also utilized for value analysis. ALPS also utilizes techniques such as customer value chain analysis, morphological analysis, scenario graphing, Pugh Concept Selection, the object process methodology, and quality function deployment. Students are asked to plan, prototype, demonstrate, and communicate the value and suitability of their solution. Projects have included designing solutions for senior mobility, designing a portable solar-powered refrigerator for vaccine storage in a disaster scenario, and a business risk mitigation plan.

Saunders-Smits et al. [10] present strategies for project-based learning implemented at Delft University of Technology in the Netherlands. Delft, a large engineering program with approximately 400 students entering each year, has incorporated five "themed" semesters into their 3 year bachelor's program in aerospace engineering. Each of the semesters incorporates a curriculum-integrated project. These projects increase in student responsibility, complexity, and scope as students progress in the program. Each project has a simulated client, places the students in a "professional" role, and is designed to be authentic. Projects during the first 2 years are run largely by teaching assistants; projects in the final year are run by Ph.D. students or junior staff members. Saunders-Smits et al. note that these projects are well received by students with 85 % indicating that the project "contributed to a better understanding of design," 91 % indicating that it "contributed a lot to their competence in working in teams," and a 68 % indicating that it provided them with a "better understanding of the relationships between the different disciplines in aerospace engineering."

Fruchter [11] describes a collaborative architecture, engineering, and construction curriculum that is a collaboration between Stanford University and five universities spanning Europe, Japan, and America. This program seeks to address fragmentation, discipline-specific education practices, and the need to advance assessment techniques for cross-disciplinary studies. Information technology is seen as a solution to this problem, facilitating collaborative cross-site and interdisciplinary work. The approach incorporated both synchronous and asynchronous elements to facilitate PBL. Fruchter suggests that the P in their use of PBL stands for problem, project, product, process, and people. It is "problem based, project organized," it produces "a product for client," and it incorporates a "reengineered process that brings people from multiple disciplines together." The goal of the work was to create an industry-analog project, taking two academic quarters, that provides a design experience, facilitates teamwork skill development, and creates "discourse that requires the constructing meanings of concept and uses of skills." The Internet was used heavily during this process to facilitate communication, document sharing, and collaborative of use of CAD Software. Fruchter proffers that, during the course, students progressed from "islands of knowledge" that understood their own discipline and that have limited knowledge about other disciplines to individuals with awareness, appreciation, and understanding of other disciplines (who understand their needs and language). The information technology solution facilitated a transition from "passive to engaged learners" and facilitated mastery and the incorporation of real-world projects.

Hall et al. [12] discuss the incorporation of active learning techniques in the unified engineering course in the Massachusetts Institute of Technology Aeronautics and Astronautics program. The unified engineering course, a key part of the sophomore curriculum in the aeronautics and astronautics department, is a lecture-based course that the department sees as its "flagship course." Active learning, in the lecture context, primarily involved providing additional opportunities for student feedback and instructor response to student feedback. The way these were implemented in the course included concept tests, classmate discussions, and "muddiest point" response cards. These mechanisms provided instructors with immediate feedback, in near real time, on student performance, areas of lack of understanding where clarification was needed. Hall et al. noted that student response to the active learning techniques was generally positive. In particular, survey results showed that students felt that professors were concerned about them and their performance due to the incorporation of these techniques.

Fevig et al. [9] presented work on an approach features an iterative engineering learning model consisting of six steps. These are lecture, recall, recall feedback, application, application feedback, and synthesis. This approach offers the benefit of higher level (from a Bloom's Taxonomy perspective) learning, while students are introduced to material that they are unfamiliar with (thus requiring a significant lecture component for material introduction). The incorporation of faculty goal-driven research into a student-centered course context is discussed. Four specific examples, drawn from graduate courses in the University of North Dakota Space Studies program, are presented: a space mission design course, the second course in a survey series, a spacecraft systems engineering course, and an orbital mechanics course. In each, students are involved in a project that causes the internalization of the materials presented. It is noted that these activities provide "a critical bridge between coursework, research, and workforce entry."

In all of the foregoing examples of PBL, and in many others, students are involved in hands-on projects that have been shown to be an effective instructional tool at all levels of education: from collegiate graduate level to primary school level [2, 5–7, 12, 13]. PBL has also been demonstrated across a wide variety of subject disciplines, including project management [14], psychology [15], physics [16], computer science [17, 18], mathematics [19], engineering entrepreneurship [20], and aerospace [10, 21], computer [22], electrical [23, 24], and mechanical [25, 26] engineering.

In addition to teaching subject-specific skills, PBL projects can teach students how to work with those outside their specific discipline, as is required in the vast majority [27] of workplaces. Gaining a shared prior knowledge base (such as through PBL techniques) can improve team efficiency [27]. Workers with interdisciplinary skills are in demand [28]; PBL also provides students with an opportunity to learn "soft" skills which are required for workplace success [29].

PBL has also been shown to have benefits that extend beyond the instruction of particular material. It has been shown to have a beneficial impact on student motivation [30], self-image and creativity [31], and material retention [32]. Field-based/realistic-environment PBL has been shown to increase students' understanding of course materials [33]. Nagda et al. [34] show that one type of PBL, research participation, can also improve student retention, particularly for at-risk students.

The benefits of PBL to student placement, after graduation, have been demonstrated by Hotaling et al. [35], and Fasse et al. [36]. Gilmore [37] even argues that techniques such as PBL, for teaching STEM disciplines, are critical to national prosperity.

In aerospace engineering and related disciplines, many students are gaining practical experience working on small spacecraft and high altitude ballooning projects. The utility of PBL for teaching aerospace engineering [2, 38], software development for aerospace applications [39], and providing other benefits [40, 41] has been demonstrated. CubeSat projects have been demonstrated to be an effective pedagogical approach [42–44].

The level of the aforementioned benefits, Zydney et al. [45] proffer, increases with the duration of participation. However, not all students reach these higher levels of benefit. While numerous reasons for premature termination of student participation in a research project exist, certain student-involved characteristics create particular risk. The subsequent section provides an overview of this topic.

8.2 Student Involvement, Faculty Research, and Risk

From the foregoing, it is clear that student participants can derive great benefit from PBL activities. They also, as part of their educational participation, can make a valuable contribution to the project that they are working on.

Faculty perceptions of student involvement in research projects are quite informative in this regard. Zydney et al. [46] proffer that faculty see students' participation as valuable, with over half of them indicating that students' contribution to their work was "important" or "very important."

While students may gain (possibly even enhanced) benefit from risk actualization, the project stands to suffer. Thus, the failure of a student to make progress is a risk that may be comparable to causing damage or other types of impact on prior work.

While student participation is valuable to faculty, it appears that project completion may be less important to students, as Prince et al. [47] demonstrated a lack of correlation between the research productivity level of faculty and students' educational benefits. The SQUIRM framework [48] was created, initially, to assess the risks applicable to student involvement in a small spacecraft project; however, it is useful for many applications beyond this.

8.3 Determining Whether the Project Will Be an Educational Program or a Program with Educational Benefits

A distinction must be made, for the purposes of goal setting and assessment, between a program enacted primarily for educational purposes and a program that offers secondary or ancillary educational benefits. Of course, a third possibility of the two being equally weighted is foreseeable; however, this may not be desirable.

8.3.1 Educational Programs

Educational programs are defined as those that have a sole or primary mission objective that is educational in nature (mirroring Swartwout's definition of the 'university-class' mission [49]). For many small spacecraft programs, this mission is to provide student training opportunities. These opportunities may be in a single discipline of study or they may be cross-disciplinary. Having a defined educational goal makes it possible for the program to be more adaptable to student needs. This type of program may also fit into the academic schedule better. Perhaps problematically, the principally educational nature may reduce the likelihood of mission completion (as these objectives may take the back seat to pedagogical goals). Additionally, given this, the program may need to be funded out of university teaching (rather than research) funds. This may reduce mission risk for faculty sponsors and allow greater student leadership and risk taking.

Interestingly, though, the lack of an external goal may remove some of the impetus behind the program (and goal attainment) and deprive students of some of the workplace-analog benefits of a program with a noneducational primary objective.

8.3.2 Programs That Provide an Educational Benefit

Unlike educational programs, some programs are designed to provide educational benefits as an ancillary purposes (or they are received as an unplanned benefit). These programs have technical, scientific, or other goals and provide students (and others) with educational benefits by virtue of their participation therein. However, unlike educational programs, they may not have defined pedagogical goals or these goals may simply be broad statements of possible ancillary benefits. Additionally, these programs may not be staffed or funded to emphasize or place lessons learned into a broader framework or context. These programs have specific mission goals that may not be aligned with the educational institution's schedule. They are higher risk to faculty members (who have research goals to deliver against, irrespective of students' contributions and capabilities). These programs may also offer less flexibility for student leadership, out-of-the-box thinking, and technical risk taking. Alternately, because of these very factors, this type of program may be a closer analog to the 'real-world' workplace, providing benefits in this regard. Additionally, it may be possible to offer some of the educational benefits that would be otherwise lacking by having students enroll and receive credit from an independent or directed study course where a faculty mentor may be able to place work activities within a broader educational context.

8.3.3 Dual-Objective Programs

Given the somewhat disjoint set of benefits presented by the two foregoing approaches (educational mission and missions with educational benefits), it is tempting to attempt to have the proverbial cake and eat it too with a dual-objective mission. While the concept is laudable, its implementation is problematic. The simple problem is, if both sets of objectives can't be met, which ones are sacrificed. For a dual primary objective mission, this question is (by definition) difficult to answer (as sacrificing either makes the mission at least a partial failure).

Alternatively, if there is a clear winner (And the other is thus a clear loser), then the mission simply had a secondary or tertiary objective (the losing one) masquerading as a primary objective.

An alternate response to this question might be that the answer is more nuanced, with parts of both declared primary objectives being sacrificeable (i.e., some parts of each are higher priority than others). This, however, may be an example of an insufficiently granular objective set. Even if granularity is not the specific problem, the intertwining elements of the objectives (and their associated dependencies) can make decision making problematic under less-than-ideal circumstances.

One way to structure a dual-primary objective mission (which was alluded to in the previous section) is to separate it into two missions: one with the technical objectives and one with the educational ones. This is not to suggest that two spacecraft need to be built. Instead, some individuals (the research leads) have responsibility for technical goal attainment (and educational activities as a secondary or tertiary objective), while others (for example, the faculty mentor of an independent or directed study course) have the educational ones as their prime focus. While some give-and-take between the two would still be required, this approach places primary focus on both; however, it may lead to some friction between the two mission component groups.

8.4 Setting Educational Goals: Technical Discipline Skills

Small spacecraft programs, whether educational programs or programs with ancillary benefits, can typically provide multiple types of benefits to their participants. The OpenOrbiter program, for example, is designed to provide several classes of benefits to its participants. These include (1) providing experience in developing a spacecraft, (2) providing experience working in an industry-analog aerospace engineering environment, (3) teaching specific technical skills, (4) allowing participants to demonstrate competence in technical skills, and (5) providing the professional development benefits from participating in a project with a highly emotive and demonstrable product.

Spacecraft development projects are normally high value projects that do not facilitate the substantive involvement of students or junior (entry level) employees. Because of this, an aspirant to spacecraft development may be a substantial portion of the way into his or her career before he or she is able to actually work with real spacecraft hardware. While this may decrease project risk, it removes a key source of innovation. It also prevents students from gaining the experience that is only possible via hands-on interaction with a flight quality or prototype system. University-run small spacecraft programs allow students complete access to the flight and prototype hardware and thus the opportunity to gain this experience.

A key challenge that small spacecraft programs can potentially aid with solving, for many students (and, by extension, the companies that eventually employ them) is learning to work with and speak the vernacular of the various disciplines that must be involved in an engineering project. Students may make it all the way through their university career (and even years into their initial employment) before they are required to communicate, about technical concepts, with those outside their particular technical 'silo'. Even many small spacecraft (and other engineering/hands-on STEM projects) only provide STEM students with experience working with other STEM students (in some cases, more aptly, providing technology and engineering students with the opportunity to work with other technology and engineering students). However, it is possible to take spacecraft program activities one step further and incorporate students from business, public administration, fine arts, education, and other disciplines, in discipline-appropriate roles (e.g., project management, Web site design, and outreach).

Many small spacecraft projects highlight technical skill learning as a key educational outcome. Skills such as systems engineering require use in a project of reasonable size to be reinforced through use and refined.

Additionally, students seeking employment at the end of their college careers desire involvement with demonstrable projects that can be highlighted to potential employers as a demonstration of their skills and abilities (and particular competency as compared to 'book-learning-only' students). Small spacecraft projects provide this demonstrable experience and are an emotive project for presentation to prospective employers. Participants, in multidisciplinary programs, can also highlight their 'cross-silo' communications and working environment as a key benefit to prospective employers.

Finally, students desire to develop and document their professional skills. Milestones such as design reviews and integrator and launch provider acceptance allow students to document their participation in terms of external standards (to differentiate participation from joining an extracurricular club, for example). Giving participants titles which reflect, appropriately, their project responsibility and authority also helps ensure this benefit accrues.

8.5 Setting Educational Goals: Nontraditional Disciplines

Outside the science, technology, engineering, and math (STEM) disciplines typically involved in small spacecraft development, the definition of educational goals is more difficult to generalize. These goals will be driven by program goals and the point that the program is currently at in achieving these goals. It will also be driven by the disciplines in which participation is desired (or foreseen) and the correlation of program needs with this discipline's activities. It is, of course, worth nothing that student participants in a program need not be majors (or even minors) in the discipline whose associate they are participating in. Thus, a marketing major seeking to develop demonstrable technical competence (or taking a technical elective, which serves as the basis for participation) might participate in an engineering role.

Alternately, an engineering student seeking to develop or demonstrate management or other soft skills might participate in a less (or non)technical role.

A few example areas of participation are now listed; however, this list is by no means exhaustive.

Management/Project Management—Take team lead or general leadership role in technical or nontechnical area.
Policy—Deal with national policy impact on program and/or local political concerns.
Marketing—Get the word out, both to the press and to the university community (e.g., recruiting).
Creative Disciplines—Design aesthetic aspects of the spacecraft. Also, students could create drawings, renderings, and such and support marketing activities with prose and artwork generation.
Education (and/or various sciences)—Design, implement, and complete onboard experiments.

Of course, the foregoing is only a subset of the numerous ways that those outside the typical engineering and closely related disciplines could participate.

8.6 Setting Educational Goals: 'Soft' and Other Skills

The definition and assurance of attainment of educational goals for 'soft' skills and other categories that do not map directly to a particular set of tasks that students will perform is inherently problematic. Numerous examples of prospective benefits which have previously been enjoyed by others, in a PBL environment, exist and may be able to be attained. Examples of such benefit, as previously discussed, include increased creativity [31], motivation [30], understanding [33], knowledge retention [32], student retention [34], and self-image [31].

How, though, can the attainment of these benefits be orchestrated and how can their attainment be assessed? The exact level and mechanism of attainment of these types of benefits will vary significantly by students pre-status and project involvement. It is, thus, difficult to create a specific plan to drive increases in areas like student creativity or self-worth. Instead a general enabling paradigm can be used to facilitate (but certainly not guarantee) each benefit category's occurrence. Each of the foregoing will now be discussed, along with enabling techniques.

Creativity—Provide an environment where students must find creative solutions to problems. Teach creative problem solving techniques and recognize and reward creative solutions. Don't discourage solutions, initially, for being 'far out there'.
Motivation—Motivation can happen on its own, but it can be helped. Find motivate students and put them in leadership and other visible roles to inspire others. Remove barriers and 'stall' points that can impair motivation and help find pathways around or through blocking challenges.

Understanding — Subject material understanding can be enhanced by direct exposure and involvement. To be effective, an existing knowledge (theoretical) foundation should exist. Students must understand how their activities and prior knowledge relate and they should be able to see how their work relates to the larger picture (i.e., project or type of task).

Knowledge Retention — The additional involvement and different learning styles provided by PBL should aid knowledge retention. Appropriate material-activity association is required for this benefit to be enjoyed.

Student Retention — PBL activities can have several benefits on student retention. They can increase students' belief in their own self-efficacy (and thus their interest in continuing to invest in their education). They can increase the students' perception of the academic program and institution. They can also increase students' happiness via creating interactions with peers and opportunities to be recognized for their skills and abilities.

Self-image — PBL can increase students' self-image through demonstrating (to themselves) that they can solve complex problems and perform important tasks. This can be aided by internal and external recognition. Additionally, highlighting improvements in competency (i.e., how far the students have come) can aid in this benefit being attained.

8.7 Formative Assessment: Assessing Students' Reasons for Participating and Using This Information

As was discussed in Chap. 1, small spacecraft development activity is increasing significantly. Between 2000 and 2013, the number of manifested "university-class" spacecraft has increased from below 5 to over 35 [50]. From its initial design by Jordi Puig-Suari and Robert Twiggs in 2000 [51], the CubeSat standard (one type of small spacecraft that is gaining in popularity due to its easy-to-integrate common form factor [52]) has matured from a tool for student learning to a mechanism for conducting bona fide science [53, 54] and other work [55].

While the benefits of the form factor for missions are clear, the reasons for student involvement in the design and development of a small spacecraft are less so. In many cases, students participate and devote their skills to small spacecraft development on a voluntary basis (or at a wage level below what they could make by obtaining an off-campus job). Do these students seek to work in the space engineering field? What reasons drive those students who are studying ancillary topics? This section begins the process of assessing why students decide to participate in small spacecraft development and what benefits they hope to obtain from doing so. This is considered in more detail in subsequent sections and chapters.

To try to answer these questions, a survey was administered to returning and prospective participants in the OpenOrbiter small spacecraft development program at the University of North Dakota. This survey, which was conducted anonymously,

8.7 Formative Assessment: Assessing Students' Reasons for Participating... 139

asked students for demographic information and then asked them to characterize their reasons for participating. These surveys were given at initial meetings used for recruiting new participants and at initial meetings of project groups.

The survey respondents included both undergraduate and graduate respondents. Respondents included 19 undergraduate and four graduate students. The undergraduates consisted of one freshman, five sophomores, seven juniors, and six seniors. Note that as a largely volunteer program, the demographic makeup of participants varies significantly from semester to semester. The current focus areas also drive the breakdown of the majors of students involved.

Students were also asked whether they had previously participated in the program or not. Twelve individuals indicated previous participation, while 11 indicated that they had not participated previously. Note that the participation is nearly evenly split between returning participants and the newly joined. Past participants were asked to indicate the duration of their previous participation. Four students indicated participation for one semester, seven indicated two semesters of participation, two students indicated three semesters of participation, and two students indicated four semesters of participation. Whether students had or were planning to receive academic credit for their participation was also assessed. Eighteen students indicated that they had not participated/were not participating for academic credit (and did not plan to do so). Three indicated participation/planned participation for a course project. One indicated participation/planned participation for an independent study project and one indicated participation/planned participation for other academic credit.

The breakdown of student participants is a function of various recruiting efforts pursued by those involved in the program. A strong recruiting effort, for example, to involve freshmen in the previous year may be largely responsible for the number of sophomores indicated and the high number of individuals with two semesters of previous participation (and probably affected the number indicating one semester as well). The number of opportunities for participating for academic credit has also expanded over time. In the first two semesters (under a thematically related precursor program), there were only two students who participated for academic credit; the third semester had three participants for academic credit, and the fourth semester has allowed seven individuals to participate for academic credit. This survey was taken prior to the establishment of one of the opportunities for for-credit participation, so several of the individuals who indicated that they were not participating for academic credit ended up doing so (some others had graduated and thus didn't take the survey). Additionally to participation as part of a senior design, junior design or other whole-class participation opportunity, several students have had the opportunity to perform work on the project to satisfy a component requirement of a class. This type of participation is not included in this category.

The first seven questions collected demographic data, while subsequent questions assessed student expectations from program participation. The responses to these questions are presented in the subsequent section.

In question eight, students were asked to select all of the benefits that they hoped to gain through their participation. The list of possible areas of benefit that could be selected is presented below. Students were also given the opportunity to write in other areas of benefit. This list is based on pre-identified project goals and other benefits that students indicated they believed they had received or would like to receive through other surveys [38] and anecdotally.

Knowledge about spacecraft design	Experience working on a large group project
Knowledge about structured design processes	Experience with a structured design process
Knowledge about a particular technical topic	Experience related to a particular technical topic
Knowledge about project management	Project management experience
Knowledge about time management	Time management experience
Leadership experience	Improving leadership skills
Improving technical skills	Improving project management skills
Improving time management skills	Understanding of how my discipline relates to others
Experience working with those from other	Learn other discipline's technical disciplines details/terminology
Real-world project experience	Improved chance of being hired in desired field
Item for resume	Ability to present at professional conference
Improved presentation skills	Ability to present at professional conference
Inclusion as author on technical paper	Recognition in the university community

The responses of students to this question are summarized in Figs. 8.1, 8.2, and 8.3. Figure 8.1 presents overall counts of the number of respondents who indicated that they hoped to gain each particular area of benefit. Figure 8.2 indicates the percentage

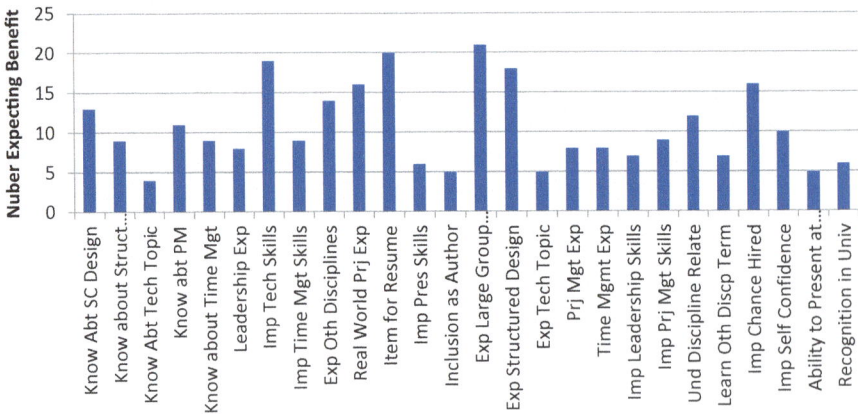

Fig. 8.1 Overview of benefits sought by participants (note that the abbreviated titles and order correspond to the above list)

8.7 Formative Assessment: Assessing Students' Reasons for Participating... 141

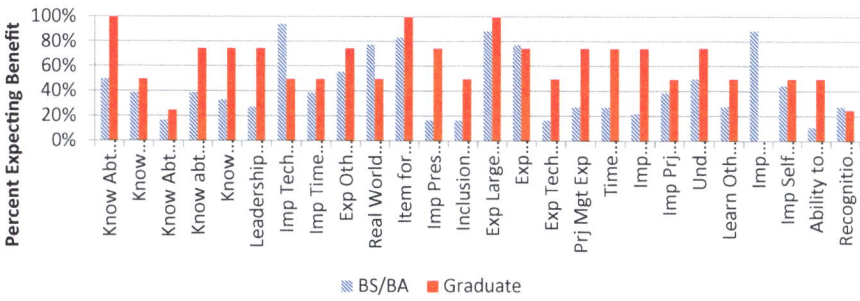

Fig. 8.2 Benefits sought by participants, by undergraduate/graduate status

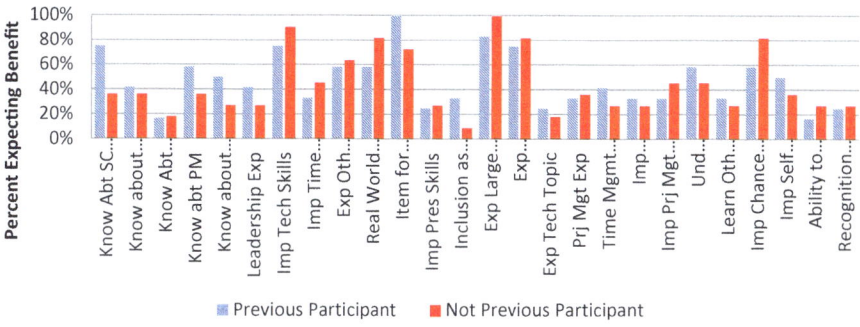

Fig. 8.3 Benefits sought by participants, by whether they have previously participated

of graduate and undergraduate students who indicated that they hoped to obtain each type of benefit. Figure 8.3 compares the responses of new entrants to those who have previously participated. Note that in most cases the expectations of previous participants and new entrants are closely correlated. This would tend to suggest that these expectations are being met. This is assessed more fully, subsequently.

In question nine, students were asked to rank their top three areas of benefit by importance. Figures 8.4 and 8.5 depict the responses to this question, with experience in a large group project, real-world project experience, and improved technical skills ranking first through third.

The next four questions sought to assess specific reasons for students joining. Questions ten and eleven asked students whether they were interested in seeking employment in the field that they were or planned to participate in and whether they believed that participation would aid them in securing employment. Both of these questions were responded to on a nine-point scale, ranging from 9-Strongly Agree to 5-Neutral to 1-Strongly Disagree. In both cases, the favorable answer (interest in seeking employment and participation aiding in securing employment) would correlate with the 9-to-5 scale range, while those believing the opposite would indicate between 5 and 1. Figures 8.6 and 8.7 present a histogram of responses to these

142   8 Setting Educational Goals and Formative Assessment

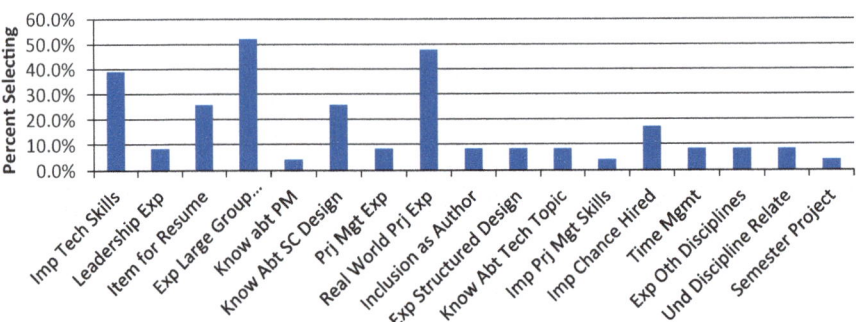

Fig. 8.4 Choices selected by respondents as one of their top three areas of desired benefit

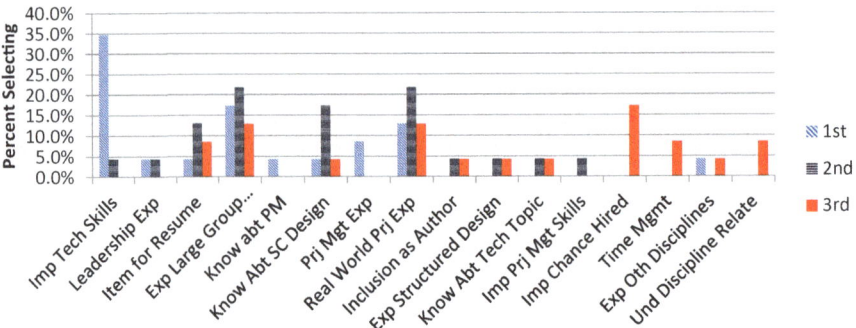

Fig. 8.5 Choices selected by respondents for each of the top three areas of desired benefit

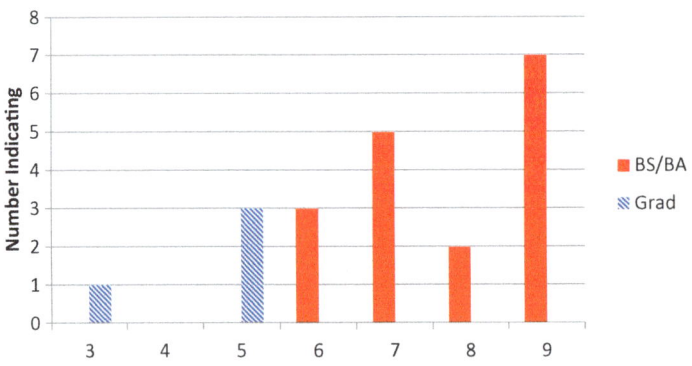

Fig. 8.6 Participant response regarding whether they are seeking employment in this field

questions. The responses of undergraduate and graduate students are compared. Note that the responses of the graduate student respondents are generally less favorable than those of the undergraduates. This may be attributable to different career aspirations, existing experience levels (and thus less perception of additional benefit to be gained), or other factors.

8.8 Formative Assessment: Assessing Whether Goals Are Being Met

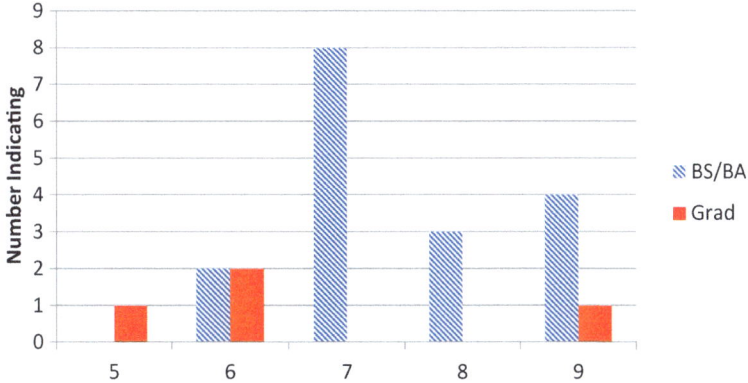

Fig. 8.7 Participant response regarding whether they believe participating will aid employment

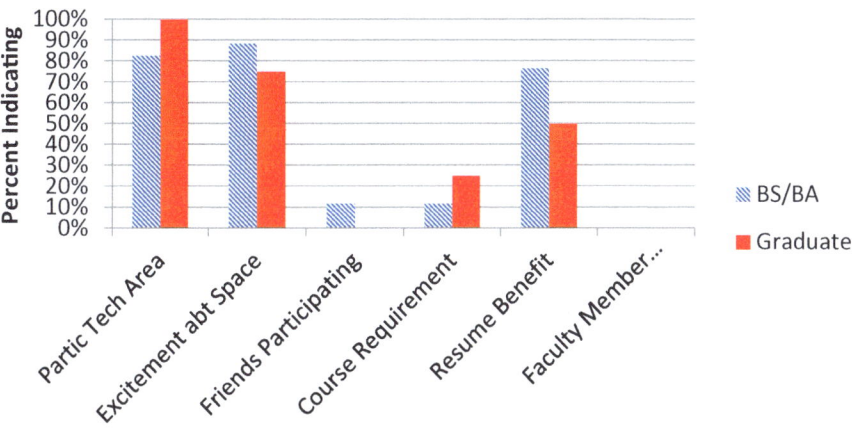

Fig. 8.8 Reason for participating, by undergraduate/graduate status

Finally, respondents were asked to indicate specific factors that drove their decision to join (in question 12). The responses to this question are presented in Figs. 8.8 and 8.9.

8.8 Formative Assessment: Assessing Whether Goals Are Being Met

Previous sections have discussed the need to assess what student participants in small spacecraft programs, particularly programs for which education is a primary objective, expected to gain from their participation. The need for this assessment is

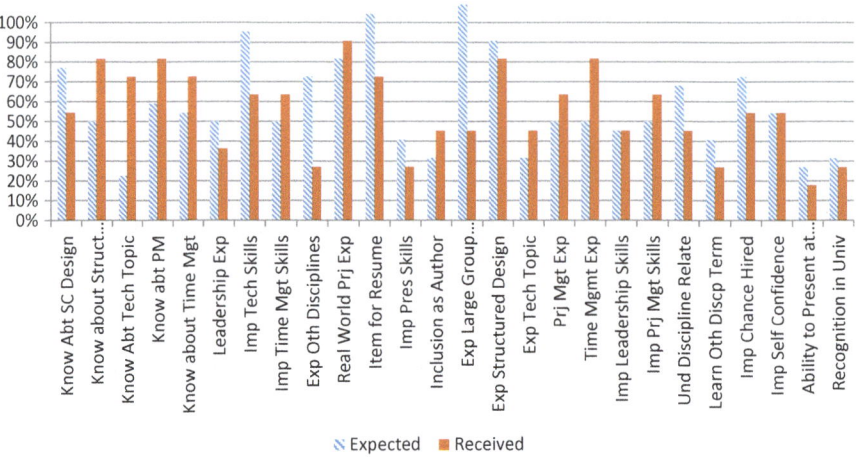

Fig. 8.9 Benefits expected and received by respondents. Source: [56]

driven by the significant increase in the creation of "university-class" spacecraft, which has grown at least sevenfold [50] between 2000 and 2013 and shows every indication of continuing to increase. To facilitate the continued growth of these programs it is important to ensure that their goals are matched to student needs.

Focus now turns to determining the extent to which student expectations are being met by small satellite program participation. This exercise is a critical formative aspect of any small satellite program. To demonstrate the process, we describe how we assessed this for student participants in the University of North Dakota's OpenOrbiter program. To this end, the areas of expected/desired benefit identified by students prior to participation are compared with the benefits attained and the correlation between the two is assessed.

The experimental design for the pre-participation survey was discussed in [4]. The post-participation survey followed this same design, except that students were asked to identify the benefits received instead of expected. To facilitate comparison, the pre-participation survey was administered at the beginning of a semester and the post-participation survey was administered at the end. As is typical with extracurricular enrichment activities, significant attrition occurred during the semester. In both cases, the survey was administered to all individuals attending the relevant meetings during the week of administration (and these attendance levels were not abnormal). With the post-participation survey, two respondents' responses have been excluded. The first was excluded due to the indicated lack of participation during the semester (making the data of little value in assessing the benefits of participation and a candidate for removal as an outlier, even without the provided explanation by the respondent). The second exclusion was attributable to a respondent personal issue.

Table 8.1 Class status of undergraduate respondents

	Pre-survey	Post-survey
Senior	6	2
Junior	7	4
Sophomore	5	3
Freshman	1	0

Survey respondents included both graduate and undergraduate students. In the pre-participation survey, 19 undergraduates and four graduate students responded. In the post-participation survey, responses were received (after exclusions) from nine undergraduate and two graduate students. The class membership of the respondents to both the pre-participation and post-participation surveys is indicated in Table 8.1. Both surveys were conducted anonymously and collected the demographic information (discussed above) in addition to data (presented in Sect. 8.4) regarding their expectations or benefits received.

In addition to being asked about their major, minor, graduate versus undergraduate status and class level, students were also asked whether they had previously participated in the OpenOrbiter program or not. In the initial survey, 12 students indicated prior participation and 11 indicated that they had not. In the post-participation survey, five students indicated that they had participated beyond the current semester and six indicated that this was their first semester of participation. Two of these post-participation respondents indicating prior participation indicated a duration of prior participation of five semesters; the remainder indicated a period of participation of two semesters. In the pre-participation survey, four students indicated one semester of participation, seven indicated two semesters, two indicated three semesters, and two indicated four semesters of participation. Of the post-participation respondents, six participated for academic credit during the semester (the pre-participation responses were gathered prior to the introduction of one form of participation for credit during the semester and thus cannot be compared to these post-participation numbers). The academic credit participants had a significantly higher retention rate (six of seven completing) as compared to the nonacademic credit participants, as might be expected.

8.8.1 Data Collected

In both surveys, respondents were asked to indicate all of the benefits that they expected (pre-participation survey) or had (post-participation survey) received in response to the eighth question. The list of possible choices presented is listed below. Respondents were also given the opportunity to write in any additional areas of benefit that they expected/hoped to receive or had received. The initial list was generated through prior surveys, such as those presented in [38], and other anecdotal feedback.

Knowledge about spacecraft design	Experience working on a large group project
Knowledge about structured design processes	Experience with a structured design process
Knowledge about a particular technical topic	Experience related to a particular technical topic
Knowledge about project management	Project management experience
Knowledge about time management	Time management experience
Leadership experience	Improving leadership skills
Improving technical skills	Improving project management skills
Improving time management skills	Understanding of how my discipline relates to others
Experience working with those from other	Learn other discipline's technical disciplines details/terminology
Real-world project experience	Improved chance of being hired in desired field
Item for resume	Ability to present at professional conference
Improved presentation skills	Ability to present at professional conference
Inclusion as author on technical paper	Recognition in the university community

In Fig. 8.9, the benefits expected and received by respondents are presented. In 13 (of the 26 categories), a greater percentage of respondents expected the benefit than received it. In two cases, the level of those expecting and receiving the benefit was the same and in all other (11) cases, a higher percentage of individuals indicated receipt of the benefit than indicated expectation of receiving it.

8.9 Incorporating the Results of Formative Assessment

A final critical consideration is how the results of formative assessment are incorporated. This, again, will depend on whether the program seeks to primarily serve educational goals (and, if so, what those goals are) or technical objectives. In the former case, specific goals that originate from course (or academic program, etc.) requirements and development progression needs may need to be blended (to some extent) with participant expectations and desires. For those participating outside of an academic course framework, program progression needs can be balanced with participant expectations and desires. For programs that primarily focus on technical objectives, the educational benefits (as well as the formative feedback provided with regards to the desires and expectations of participants from involvement) may take a second seat to program technical development and timeframe needs. In any case, the incorporation of the desires and expectations of program participants will help to ensure their continued interest in and excitement regarding the program. Whether participants are volunteers or paid staff, longevity of participation, and excitement are critical to program success. Thus, taking the maximum possible notice of participant desires and expectations and incorporating them as possible can only be beneficial to the program.

8.10 Conclusion

This chapter has focused on the process of forming the educational components of a small spacecraft program. It has discussed two types of programs (those with primarily educational objectives and those with primarily technical/scientific objectives) as well as different prospective hybrids between the two. Then, focus turned to determining what participants are interested in getting from their participation (in the form of technical skills and abilities, soft skills, and other benefits) and how this can be incorporated in the program design. The assessment of whether these benefits were being attained was also discussed. Subsequent chapters will continue this discussion with a focus on summative assessment as well as the presentation of the results of prior program assessment.

References

1. Straub, J., C. Korvald, A. Nervold, A. Mohammad, N. Root, N. Long, and D. Torgerson. 2013. OpenOrbiter: A low-cost, educational prototype CubeSat mission architecture. *Machines* 1: 1–32.
2. Straub, J., J. Berk, A. Nervold, and D. Whalen. 2013. OpenOrbiter: An interdisciplinary, student run space program. *Advances in Education* 2: 4–10.
3. Straub, J. 2014. Extending the student qualitative undertaking involvement risk model. *Journal of Aerospace Technology and Management* 6(3): 333–352.
4. Straub, J., and D. Whalen. 2013. Student expectations from participating in a small spacecraft development program. *Aerospace* 1(1): 18–30.
5. Mathers, N., A. Goktogen, J. Rankin, and M. Anderson. 2012. Robotic mission to mars: Hands-on, minds-on, web-based learning. *Acta Astronautica* 80: 124–131.
6. Mountrakis, G., and D. Triantakonstantis. 2012. Inquiry-based learning in remote sensing: A space balloon educational experiment. *Journal of Geography in Higher Education* 36(3): 385–401.
7. Brodeur, D.R., P.W. Young, and K.B. Blair. 2002. Problem-based learning in aerospace engineering education. Presented at Proceedings of the 2002 American Society for Engineering Education Annual Conference and Exposition.
8. Haruyama, S., S. Kim, K. Beiter, G. Dijkema, and O. de Weck. 2013. A new project-based curriculum of design thinking with systems engineering techniques. *International Journal of System of Systems Engineering* 4.
9. Fevig, R., J. Casler, and J. Straub. 2012. Blending research and teaching through near-earth asteroid resource assessment. Presented at Space Resources Roundtable and Planetary & Terrestrial Mining Sciences Symposium..
10. Saunders-Smits, G.N., P. Roling, V. Brügemann, N. Timmer, and J. Melkert. 2012. Using the engineering design cycle to develop integrated project based learning in aerospace engineering. Presented at International Conference on Innovation, Practice and Research in Engineering Education.
11. Fruchter, R. 1999. A/E/C teamwork: A collaborative design and learning space. *Journal of Computing in Civil Engineering* 13(4): 261–269.
12. Hall, S.R., I. Waitz, D.R. Brodeur, D.H. Soderholm, and R. Nasr. 2002. Adoption of active learning in a lecture-based engineering class. Presented at Proceedings of the 32nd Annual Frontiers in Education Conference.

13. Nordlie J., and R. Fevig. 2011. Blending research and teaching through high-altitude balloon projects. In *Proceedings of the 2nd Annual Academic High Altitude Conference, Ames, IA*.
14. Pollard, C.E. 2012. Lessons learned from client projects in an undergraduate project management course. *Journal of Information Systems Education* 23(3): 271–282.
15. Dahlgren, M.A., and L. Dahlgren. 2002. Portraits of PBL: Students' experiences of the characteristics of problem-based learning in physiotherapy, computer engineering and psychology. *Instructional Science* 30(2): 111–127.
16. Duch, B.J. 1996. Problem-based learning in physics: The power of students teaching students. *Journal of College Science Teaching* 15(5): 326–329.
17. Broman, D., K. Sandahl, and M. Abu Baker. 2012. The company approach to software engineering project courses. *IEEE Transactions on Education* 55(4): 445–452. doi:10.1109/TE.2012.2187208.
18. Correll, N., R. Wing, and D. Coleman. 2013. A one-year introductory robotics curriculum for computer science upperclassmen. *IEEE Transactions on Education* 56(1): 54–60. doi:10.1109/TE.2012.2220774.
19. Roh, K.H. 2003. Problem-based learning in mathematics. *ERIC Clearing House for Science Mathematics and Environmental Education* pp. 2004–2003.
20. Okudan, G.E., and S.E. Rzasa. 2006. A project-based approach to entrepreneurial leadership education. *Technovation* 26(2): 195–210.
21. Jayaram, S., L. Boyer, J. George, K. Ravindra, and K. Mitchell. 2010. Project-based introduction to aerospace engineering course: A model rocket. *Acta Astronautica* 66(9): 1525–1533.
22. Qidwai, U. 2011. Fun to learn: Project-based learning in robotics for computer engineers. *ACM Inroads* 2(1): 42–45.
23. Bütün, E. 2005. Teaching genetic algorithms in electrical engineering education: A problem-based learning approach. *International Journal of Electrical Engineering Education* 42(3): 223–233.
24. Roberto de Camargo Ribeiro, L. 2008. Electrical engineering students evaluate problem-based learning (PBL). *International Journal of Electrical Engineering Education* 45(2): 152–161.
25. Coller, B.D., and M.J. Scott. 2009. Effectiveness of using a video game to teach a course in mechanical engineering. *Computers & Education* 53(3): 900–912.
26. Robson, N., I.S. Dalmis, and V. Trenev. 2012. Discovery learning in mechanical engineering design: Case-based learning or learning by exploring? Presented at 2012 ASEE Annual Conference.
27. Hayne, S., M. Pendergast, and C. Smith. 2012. Team composition, knowledge and collaboration. Presented at SAIS 2012 Proceedings.
28. Sulaiman, S.A., H.M. Git, and A.M.A. Rani. 2010. Engineering team project course for development of teamwork skills. Presented at Proceedings of the 3rd Regional Conference on Engineering Education and Research in Higher Education.
29. Jackson, D., and P. Hancock. 2010. Non-technical skills in undergraduate degrees in business: Development and transfer. *Education Research and Perspectives* 37(1): 52–84.
30. Doppelt, Y. 2003. Implementation and assessment of project-based learning in a flexible environment. *International Journal of Technology and Design Education* 13(3): 255–272.
31. Ayob, A., R.A. Majid, A. Hussain, and M.M. Mustaffa. 2012. Creativity enhancement through experiential learning. *Advances in Natural and Applied Science* 6(2): 94–99.
32. Bauerle, T.L., and T.D. Park. 2012. Experiential learning enhances student knowledge retention in the plant sciences. *HortTechnology* 22(5): 715–718.
33. Simons, L., L. Fehr, N. Blank, H. Connell, D. Georganas, D. Fernandez, and V. Peterson. 2012. Lessons learned from experiential learning: What do students learn from a practicum/internship? *International Journal of Teaching and Learning in Higher Education* 24(3): 325–334.
34. Nagda, B.A., S.R. Gregerman, J. Jonides, W. von Hippel, and J.S. Lerner. 1998. Undergraduate student-faculty research partnerships affect student retention. *The Review of Higher Education* 22(1): 55–72.

35. Hotaling, N., B.B. Fasse, L.F. Bost, C.D. Hermann, and C.R. Forest. 2012. A quantitative analysis of the effects of a multidisciplinary engineering capstone design course. *Journal of Engineering Education* 101(4): 630–656.
36. Fasse, B., N. Hotaling, C. Forest, L. Bost, and C. Hermann. 2012. The case for multidisciplinary capstone design: A quantitative analysis of impact on job placement and product quality. Presented at Proceedings of the Biomedical Engineering Society (BMES) 2012 Annual Meeting. http://pbl.gatech.edu/wp-content/uploads/2013/05/fasse_bmes_2012.pdf.
37. Gilmore, M. 2013. Improvement of STEM education: Experiential learning is the key. *Modern Chemistry and Applications* 1: e109.
38. Straub, J., and D. Whalen. 2013. An assessment of educational benefits from the OpenOrbiter space program. *Education Sciences* 3(3): 259–278.
39. Straub, J., D. Whalen, and R. Marsh. 2014. Assessing the value of the OpenOrbiter program's research experience for undergraduates. *Sage OPEN* 2014.
40. Swartwout, M. 2004. University-class satellites: From marginal utility to 'disruptive' research platforms. Presented at 18th Annual AIAA/USU Conference on Small Satellites, 11pp.
41. ———. 2011. AC 2011-1151: Significance of student-built spacecraft design programs it's impact on spacecraft engineering education over the last ten years. Presented at Proceedings of the American Society for Engineering Education Annual Conference. http://www.asee.org/file_server/papers/attachment/file/0001/1307/paper-final.pdf.
42. Larsen, J.A., and J.D. Nielsen. 2011. Development of cubesats in an educational context. Presented at Recent Advances in Space Technologies (RAST), 2011 5th International Conference On.
43. Larsen, J.A., J.F.D. Nielsen, and C. Zhou. 2013. Motivating students to develop satellites in problem and project-based learning (PBL) environment. *International Journal of Engineering Pedagogy (iJEP)* 3(3): 11–17.
44. Straub, J. 2013. OpenOrbiter: Analysis of a student-run space program. Presented at Proceedings of the 64th International Astronautical Congress.
45. Zydney, A.L., J.S. Bennett, A. Shahid, and K.W. Bauer. 2002. Impact of undergraduate research experience in engineering. *Journal of Engineering Education* 91(2): 151–157.
46. ———. 2002. Faculty perspectives regarding the undergraduate research experience in science and engineering. *Journal of Engineering Education* 91(3): 291–297.
47. Prince, M.J., R.M. Felder, and R. Brent. 2007. Does faculty research improve undergraduate teaching? An analysis of existing and potential synergies. *Journal of Engineering Education* 96(4): 283–294.
48. Straub, J., R. Fevig, J. Casler, and O. Yadav. 2013. Risk analysis & management in student-centered spacecraft development projects. In *Proceedings of the 2013 Reliability and Maintainability Symposium, Orlando, FL, USA*.
49. Swartwout, M. 2006. Twenty (plus) years of university-class spacecraft: A review of what was, an understanding of what is, and a look at what should be next.
50. ———. 2013. The long-threatened flood of university-class spacecraft (and CubeSats) has come: Analyzing the numbers. Presented at Proceedings of the 27th Annual AIAA/USU Conference on Small Satellites.
51. Twiggs, R., and B. Malphrus. 2011. CubeSats. In *Space Mission Engineering: The New SMAD*, ed. J.R. Wertz, D.F. Everett, and J.J. Puschell, 803–821. Hawthorne, CA: Microcosm Press.
52. California Polytechnic State University. 2009. *CubeSat design specification, revision 12*. California Polytechnic State University, San Luis Obispo, California, August 1, 2009.
53. Muylaert, J., R. Reinhard, C. Asma, J. Buchlin, P. Rambaud, and M. Vetrano. 2009. QB50: An international network of 50 cubesats for multi-point, in-situ measurements in the lower thermosphere and for re-entry research. Presented at ESA Atmospheric Science Conference, Barcelona, Spain.

54. Chirayath, V., and B. Mahlstedt. 2012. HiMARC 3D-high-speed, multispectral, adaptive resolution stereographic CubeSat imaging constellation. Presented at Proceedings of the AIAA/USU 2012 Small Satellite Conference.
55. Weeks, D., A.B. Marley, and J. London III. 2009. SMDC-ONE: An army nanosatellite technology demonstration. Presented at SMDC-ONE: An Army Nanosatellite Technology Demonstration.
56. Straub, J., and R. Marsh. 2014. Assessment of educational expectations, outcomes and benefits from Small Satellite Program participation. Proceedings of the 2014 AIAA/USU Conference on Small Satellites. Logan, UT.

Chapter 9
Summative Assessment

This chapter discusses the assessment of small spacecraft development programs and why this assessment is important. It begins with a brief overview of the concept of project-based learning (PBL), summarizing a subset of the material presented in previous chapters. It then discusses the process of determining program value, describing a summative assessment approach, and survey instruments that can aid in the summative assessment process. The process of tracking value over time is then discussed, followed by a discussion of reporting the value of the program. Finally, the importance of being able to explain the utility and value of the program is discussed, before concluding.

9.1 Overview

Student learning in many small spacecraft programs is premised on the concept of learning by doing. While some material may be formally taught through traditional lectures and coursework and other material learned in a conventional academic manner through textbooks, a significant portion of the value of the program to participants is allowing them to make decisions, try conventional and new things, and witness the results first hand.

This chapter is based on, revises and extends the paper "OpenOrbiter: Analysis of a Student-Run Space Program" [1], *"How We're Changing Computer Science Education and How You Can Help"* [2], *"Evaluation of the Educational Impact of Participation Time in a Small Spacecraft Development Program"* [3] *and "Initial Results from the First National Survey of Student Outcomes from Small Satellite Program Participation"* [4].

9.2 Background

Three areas of background are relevant to this chapter's focus on summative assessment. First, prior work on PBL and experiential education (EE), the educational approaches used by most small spacecraft programs, is presented. Then, an overview of small satellites is provided. Finally, background on assessment is discussed.

9.2.1 Project-Based Learning and Experiential Education

Of course, the concept of learning by doing is certainly not new; apprenticeships have been utilized, historically, as a training mechanism [5, 6]. PBL and EE use this concept within the modern-day formal education system. They have been shown to be effective at all levels of education: from primary to university level [7–13] to adult level [14, 15] education. Its utility in a multitude of disciplines has also been shown. These include electrical [16, 17], mechanical [18–20], computer [21], and aerospace [22, 23] engineering, computer science [24, 25], engineering entrepreneurship [26], project management [27], mathematics [28], and physics [29]. Doppelt [30] has demonstrated the benefits of this approach with regards to student motivation and self-image. Ayob et al. [31] demonstrated PBL and EE's positive effect on student creativity. PBL and EE have also been shown to positively affect subject material understanding [32], knowledge retention [33], students' retention in an academic program [34], workforce preparedness [33], and employment placement [35]. Hotaling et al. [36] and Fasse et al. [37] have even shown the positive impact of the use of PBL and EE on student placement following graduation. Perhaps most notably, Gilmore [38] goes as far as to contend that STEM education will determine the future viability of nations, suggesting that PBL and EE are critical to the United States' ability to compete globally. The utility of PBL and EE in the context of small spacecraft development has also been previously demonstrated [7, 39, 40]. In small spacecraft programs, students are able to learn a specific spacecraft design and development process (e.g., [41, 42]) in addition to gaining valuable experience in the application of formal processes in general, design decision making, and other discipline-specific technical skills.

In the university context, PBL can occur in several formats. Students may engage in PBL activities as part of a regular course, such as a course project [7] and a PBL-style course. They may participate as part of an independent or directed study [43] or to satisfy a senior design requirement [44]. They might also participate for extracurricular educational enrichment [43].

9.2.2 Small Spacecraft Development

Small spacecraft come in many varieties. In fact, the exact definition of what is a small spacecraft is elusive. Prefixes have been defined [42] to classify types of spacecraft; however, there is no line defining where small ends and larger sizes

9.2 Background

begin. Swartwout [45, 46] proffers that size isn't the defining attribute. Instead, he suggests that the so-called university-class spacecraft should be defined by their educational missions, risk tolerance, and the ability to serve as a testing bead for out-of-the-box concepts. The CubeSat is one form factor that is commonly used for university-class spacecraft. Developed initially by Bob Twiggs and Jordi Puig-Suari as a tool to facilitate aerospace engineering education [47], CubeSats are now widely used by education [48, 49] as well as being developed for science [50–53], government [54], military [55, 56], and commercial [57, 58] purposes. Their development is being aided [59] by the availability of free-to-developer launch services from the U.S. Air Force [60], NASA [61], and the ESA [62]. Lower cost commercial launches are also on the horizon [63, 64]. Low cost development approaches, such as OPEN [65], are also enabling adoption via reducing the cost of spacecraft development. In 2013, 30 academic and 50 nonacademic CubeSats were manifested and over 100 institutions had participated in the development of a CubeSat-class spacecraft [49].

From limited beginnings as a tool for aerospace education instruction [47, 66], CubeSats use has grown significantly. In addition to educational institutions, whose goals for spacecraft may be education related or pushing technical boundaries [46], CubeSats are now being used for bona fide research [51, 52, 67] and by military [55, 56, 68], government [69], and commercial [57, 58] entities. In addition to their initial 1-U (10 cm × 10 cm × 11 cm, 1.33 kg) configuration [70], standards have been or are being developed for multiples of this, such as 2-U, 3-U, 6-U, and larger [71].

CubeSat development is being aided by the availability of low cost launches from commercial providers [63, 64] and free-to-developer launch services to qualified institutions in both the United States [61] and in ESA member states [62]. Initiatives such as OPEN are also reducing the costs [65] of spacecraft development in various ways. The presence of numerous commercial providers with space-qualified hardware makes entry easy for governmental, commercial, and other users looking for a closer-to-turnkey solution.

9.2.3 Assessment

Evaluating the performance of courses, education approaches, and educators is a subject that provokes no shortage of problems. Significant disagreement exists regarding how to achieve the best results for students, or even what results should be generated and assessed [72–74]. Others fear that determining an evaluative criterion may allow an administration to 'clean house' of those not subscribing to an approach. O'Mahony and Garavan [75] contend that this "managerialism" perception, the notion that university leadership uses systems to manage the school in a business-like way and seeks to "advance strategic objectives" for some can be problematic. However, a robust approach, which considers knowledge, skill, and experience attainment may identify numerous benefits and the trades that must be made to get each. This may include benefits beyond what are typically assessed, such as enhanced creativity [31], motivation and self-image [30], and job placement benefits [36].

Current policy makers perceive an ever-growing cost of higher education [76] with generally positive results, but which suffers from a difficulty of deconfounding the selection effect (of who seeks to attend and is admitted to colleges) from the impact of the college's educational services [77]. Baum, Kurose, and McPherson [77] proffer that value is being created; however, its characterization in the specific is elusive—even though student earnings differentials [77, 78] demonstrate the presence of significant value. Many metrics show US education systems trailing behind other countries, across all levels (e.g., [79, 80]). However, these measures may exclude metrics (such as the hands-on experience generated by project-based [81] and other experiential education techniques [82, 83]) under which the United States may perform more favorably. Alston et al. [84] indicate that many of these other skills are key indicators of students' ability to succeed in the workplace. The evaluation of PBL/EE, particularly in a group project context, is inherently problematic, as each student's educational focus and benefits are inherently different. Subsequent sections discuss prospective approaches to this assessment challenge.

9.3 Small Spacecraft Programs and Their Goals

Many small spacecraft development programs have distinct student participant-related goals. For OOSDI, for example, identified goals included allowing student participants to develop and demonstrate specific technical skills, time and project management skills, presentation skills and comfort giving presentations, leadership skills and experience, and an understanding of how to work with those in other disciplines [43]. As was discussed in Chap. 8, the technical skill category is perhaps the most straightforward. Many students participate in the area of their academic major (or a closely related area). Some opt to gain experience and knowledge in an entirely different area that interests them. In either case, specific topics of learning can be selected for assessment. In addition, all participants learn about the spacecraft design and development process and unique spacecraft development considerations.

Time and project management skills are also an identified area of focus for OOSDI [3, 43]. The large scale and level of involvement in the project facilitates the application and appreciation of the importance of these techniques in a way not possible in many smaller projects (where, simply by virtue of the closeness of group members, good results may occur in spite of poor management).

Time management learning, while certainly possible in the context of smaller projects, is aided through its use in a large small spacecraft development project. The importance of delivering what is promised on time, when others are waiting for it, creates a particular impetus helpful for learning best practices that might be otherwise lacking.

For OOSDI, to aid project management, a variety of tools were used. Software development groups learned to use source code management tools and techniques [86]. Other groups have benefited from project management software and online collaboration tools.

Presentation skills and comfort are another area where small spacecraft programs can provide value. For OOSDI, they were identified at the outset as another area of focus. Due to this, numerous presentations have been made including a significant number with undergraduate student first authors (see [87–97]). Opportunities for presentation skill learning, use, and improvement have also occurred in regularly scheduled group meetings where student participants must present the results of their weekly efforts to other members of their group.

Leadership skill development is another key area of benefit. However, this was not initially identified as a core focus of OOSDI (though it was explicitly acknowledged as a benefit) and was not assessed during initial program assessments. However, due to student feedback regarding it being a reason for and a valuable benefit gained from participation [43], it was added as a focus area and has been subsequently assessed.

Gaining experience in working with those in other disciplines is another prospective area of benefit. Student participants can learn about the technical, logistical, workplace, and other needs of practitioners in other disciplines. They can also learn the terminology used in these disciplines. As most students, upon entering the workforce, will be required to interact with those from different disciplines (e.g., managers, subordinates, coworkers) this experience prepares them for the 'real world' and gives them an advantage (both during the hiring process and initial work periods) as compared to graduates without this experience.

Preliminary work on the assessment of student expectations and desires from small spacecraft program participation indicates that at least 26 different prospective sources of perceived benefit exist [1]. This topic will be discussed more fully in Chap. 10.

9.4 Determining Program Value

The dramatic increase in CubeSat development in academia [48, 49] has driven a need to make sure that academic CubeSat programs are adequately preparing students for the challenges awaiting them upon graduation. Educational benefit assessment was a key pillar of the OOSDI, as it demonstrates the value for student participants at the University of North Dakota as well as provides data that can be used by others to justify the creation of a small spacecraft (CubeSat) development program at their own campus. This is highly aligned with the goals of the Open Prototype for Educational NanoSats (OPEN) which seeks to reduce the costs of CubeSat development from between $50,000 and $250,000 [59] to requiring a parts cost budget of just $5000 (excluding mission-specific payload components) [65]. OPEN is doing this by making the designs, documentation, fabrication instructions, software, test plan, and other materials for its form factor maximizing design [98] available via the Internet for use by students, faculty, and researchers worldwide. OpenOrbiter is demonstrating the space readiness of the OPEN design through the fabrication and launch of an OPEN-class 1-U (10 cm × 10 cm × 11 cm) CubeSat. Methods for determining program value are now discussed.

9.4.1 Context of the Summative Assessment Process

The PBL method was used extensively in all aspects of student participation in the OpenOrbiter project, which serves as an example of the use of summative evaluation. In the context of students participating for their personal enrichment, self-guided PBL [99] occurred as students chose areas of the project to work on and selected tasks to perform meeting their personal interests and learning goals. It is important to note that the expectation surveys [100] indicated that students were participating in response to specific educational or career development goals and, generally, not due to the participation of friends or preferred faculty members. Students were, thus, in an ideal position to select a direction for their own learning and they could and did seek out project leadership and faculty mentor guidance, as required, in meeting these goals, in many cases. Students participating for independent study credit were similarly able to define their area of focus, subject to faculty mentor approval.

In most other cases, the PBL-style learning was either partially self/partially instructor-led or instructor-led [99]. Students participating in the context of the CSCI 297 class were given specific objectives and deliverables, though they were given great latitude in terms of how they planned for and reached these deliverable goals. They were also able to select the topic for their final reports from a list of project management focus areas. Students participating for class project credit and capstone credit had more flexibility, as they were able to select their focus area (to some extent, given class constraints and current project areas' status); however, the types of deliverables that they were required to produce and when they were required to be produced or presented were set by the instructor of the course or capstone/ senior project supervisor.

Limited non-PBL methods were also used at various points in the project. Several impromptu lectures on orbital mechanics or spacecraft-specific issues have been given to assist students in grasping these topics. Students in the CSCI 297 course also participated in instructor-led discussions and were required to provide critical feedback and questions related to their classmates presentations or comments. A very limited amount of formal lecturing was also used in CSCI 297 to set the path for the students' exploration of the project management concepts in an experiential manner.

All of the aforementioned learning activities were assessed using techniques described in the subsequent sections (and in Chap. 10).

9.4.2 Undergraduate Research Student Self-Assessment (URSSA) Mechanism

The Undergraduate Research Student Self-Assessment (URSSA) [101] mechanism is a highly validated [102], widely used assessment for quantifying the benefits of research participation by undergraduate students. The University of Colorado at

Boulder team that developed the URSSA conducted an 8-year-long study of undergraduate research at multiple institutions comprising over 350 interviews. They conducted three evaluations of undergraduate research programs (including an additional 350 interviews and surveying 150 students). Finally, they performed a literature review with regards to relevant studies of undergraduate research evaluation.

From this, they developed an initial survey, refined this with the so-called thinkaloud interviews to assess interpretation of the questions' wording and conducted a pilot study including more than 500 students at 24 institutions. They used confirmatory factor analysis and removed or changed items as necessary to correct remaining issues.

This instrument collects data relating to both student perception of achievement and specific outcomes. Example data from the URSSA is presented in Table 9.3 and in Chap. 10.

9.4.3 Experiment Implementation

The experimental implementation consisted of three aspects. First, at the beginning of each period of focus, student participants were surveyed to ascertain what benefits they hoped to achieve from their participation. The results of that study [100] confirmed existing plans in many areas and led to relatively minor changes (such as creating additional presentation opportunities) in others. Second, students participated in the OpenOrbiter program throughout the semester. As is typical, a number of the students participating for personal enrichment reduced or curtailed their participation at various points throughout the semester (likely in response to other demands on their time). The URSSA survey was given (without previous announcement which could result in self-selection to take/not take the survey) at the final group meetings for the semester to all of the participants who were at that meeting. Two graduate students who commenced participation as an undergraduate and continued their participation as a graduate student were included in those surveyed.

9.5 Tracking and Reporting Program Value over Time

Several reasons exist to perform the aforementioned benefit assessment. These include assessing the program for formative purposes, to attempt to maximize student benefit, and for scientific analysis purposes. Additionally, to support program longevity and long-term access to resources, it is desirable to track the value of small spacecraft programs over time. This takes two forms. First, the changing attitudes, skills, and abilities of program participants can be measured using the instruments described in Chap. 10.

Second, it is desirable to follow the progress of program participants after they exit the program (and complete their degrees) to see what impact program participation may have on them in the long term. As random assignment is not feasible, this is not a controlled study. In some areas assessment may have to rely on anecdotal evidence and it may suffer from possible confounding variables.

9.6 Reporting This Value

The prospective benefits of value tracking also require the results to be reported, in many environments. Some programs may also have specific research objectives relative to particular program aspects or approaches. The analysis of a program, in the context of securing resources, may focus, primarily, on reporting specific benefits produced by the program. This could be presented in terms of anecdotal examples, identified benefits, and/or quantitative and qualitative assessment of skill, ability, and attribute changes. Summative studies may focus both on program changes, compared to previous approaches or prior work by others, and on the associated results. They may also be more controlled, comparing the results of different approaches (within a program) to which students have been assigned. Mechanisms for the collection and analysis of quantitative (and associated qualitative) analysis, and prior work in this regard, are presented in Chap. 10. The following sections discuss techniques for reporting value, either as part of a summative assessment of local program performance or in comparison to other programs.

9.6.1 Local Reporting

Reporting the progress of the program, locally and to funding sources, will likely be necessary for most small spacecraft programs. Funders will require documentation of the results of their investment. Participants will be interested in what has been accomplished. Future participants may be interested in learning about the results of past participation when deciding whether to devote their time and effort to the project. Other stakeholders may have similar interests. Local reporting can take a variety of forms, ranging from press releases (and other communications with the media) to newsletters and social media posts to more formal analysis. A discussion of the intricacies of media relations is beyond the scope of this book. However, interested readers may wish to review the numerous books on this topic, such as Bland, Theaker and Wragg's *Effective Media Relations: How to Get Results*. Similarly, the management of social media is a rapidly evolving topic with significant information available. One source of information on this area is Richards's *Social Media: Dominating Strategies for Social Media Marketing with Twitter, Facebook, Youtube, LinkedIn, and Instagram*. Analytical local reporting can be thought of as an offshoot of assessment. A discussion of this and examples from prior work can be found in Chap. 10.

Irrespective of the format used, it is important to consider the implementation of an overarching strategy to all reporting to ensure that a consistent message is delivered. The university environment (where anyone can and is encouraged to disseminate information) can make this a challenge. Thus, developing a program culture that embodies (and is the source of) the desired message may help to reduce the prevalence of conflicting messages and confusion. Additionally, it is important to coordinate between all those responsible for various reporting activities to ensure a consistent use of terminology and to prevent conflicts in message and timing.

9.6.2 Comparison to Other Programs

In order to demonstrate the particular value of a given program, it may be desirable to compare the performance of the program to others on a national or international basis. Programs may be compared in terms of specific dimensions of interest, or generally. One approach that may be prudent is to compare the program to the national averages in terms of key areas of focus. This focus could be a particular type of skill that is developed, certain attitudes that are changed, or an ability (or preserved) or some other characteristic. To enable this type of comparison, a national (in the United States) survey of participants in CubeSat programs was conducted [4]. A subset of the results of this study is now presented. While this study certainly does not provide information on every topic that may be relevant to program assessment and comparison, it does provide a starting point and may facilitate a rapid comparative study.

The survey collected data from participants in small spacecraft development programs using Qualtrics (an online survey tool). Requests for participation were sent out via a number of channels including online mailing lists (such as the CubeSat list maintained by CalPoly) as well as mentions in several presentations. One hundred and forty-two people completed some portion of the survey; due to the length (and potentially the inapplicability of certain questions to some programs) only about one-fifth of the respondents completed the entire survey.

This survey is based on a more limited survey, previously used for analysis of the local program at the University of North Dakota, which was discussed in [3, 43, 103, 104]. As with portions of the aforementioned prior work, some of the questions were based on the Undergraduate Research Student Self-Assessment (URSSA) [101, 102]. The following sections present a subset of the results from this survey.

9.6.2.1 Information About Respondents

First, some information about the respondents is presented. Figures 9.1 and 9.2 indicate the level of involvement of the participants with the U.S. Air Force University NanoSat Program and NASA Educational Launch of NanoSats (ELaNa) Program, respectively.

Respondents were then asked to provide information about where their program is funded from. Figure 9.3 presents this information.

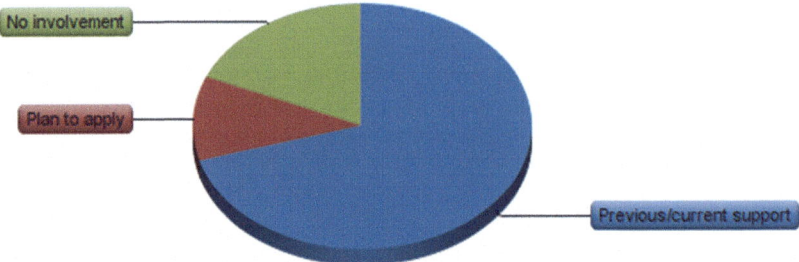

Fig. 9.1 University NanoSat program involvement

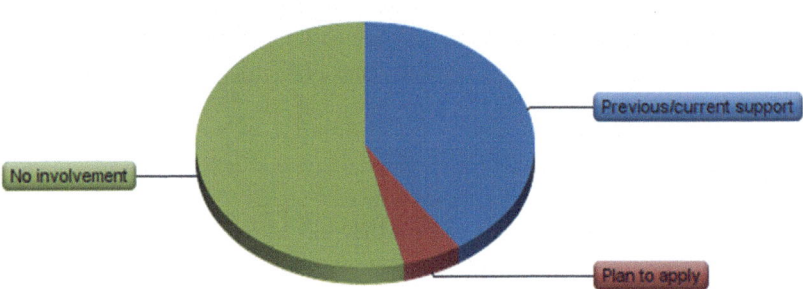

Fig. 9.2 NASA ELaNa program involvement

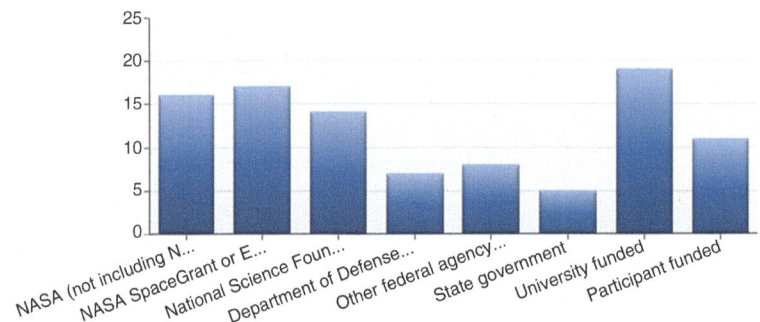

Fig. 9.3 Program funding sources

Next, information was collected about the particular individuals who were completing the survey. Figure 9.4 indicates their academic level (notably this is largely bachelor's level students). Figure 9.5 presents respondents' answer to the question of how long they have spent in their current academic program.

Undergraduate respondents were also asked to indicate their current academic year level. This is presented in Fig. 9.6. Notably, the majority of respondents were upperclassmen (seniors and juniors). Figure 9.7 indicates the GPAs reported by the respondents. Note that 90 % reported a GPA of 3.0 or higher.

9.6 Reporting This Value 161

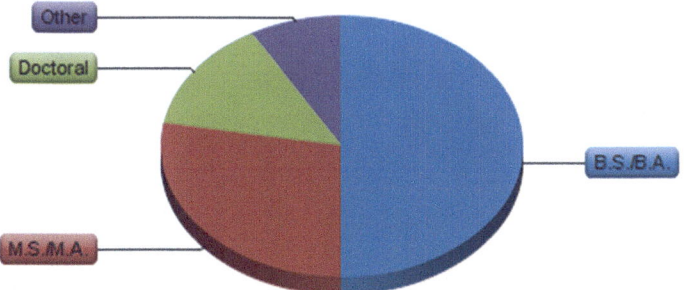

Fig. 9.4 Academic level of respondents

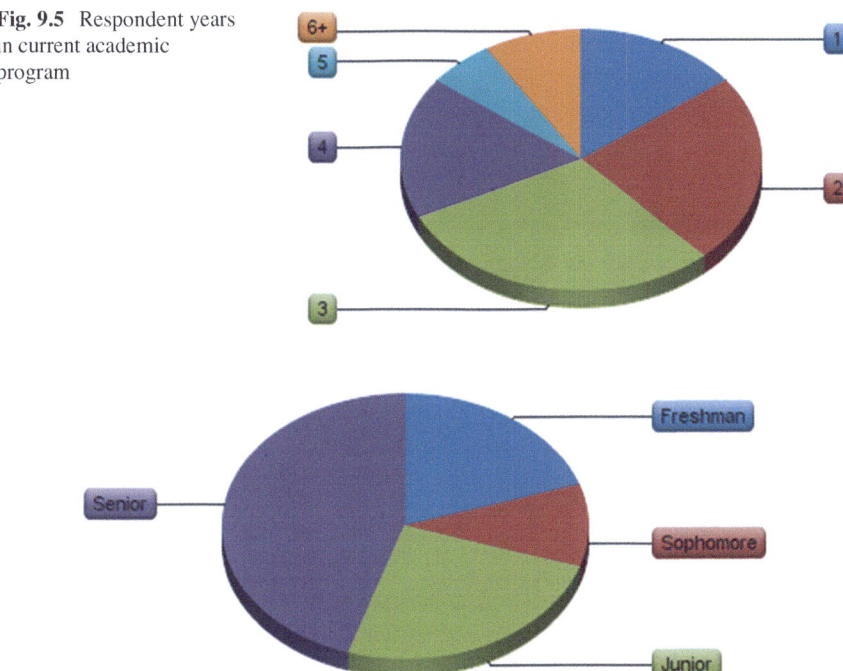

Fig. 9.5 Respondent years in current academic program

Fig. 9.6 Respondent class, for undergraduates

9.6.2.2 Respondent Participation

Respondents were also asked to indicate the nature of their participation: whether they served in a team or group leadership role or not. Responses to this question are presented in Fig. 9.8. Notably, approximately half of respondents fell into each category.

Respondents were also asked to indicate how much time each week they spent on program activities. Nearly 70 % of respondents indicated that they spent more than

162   9 Summative Assessment

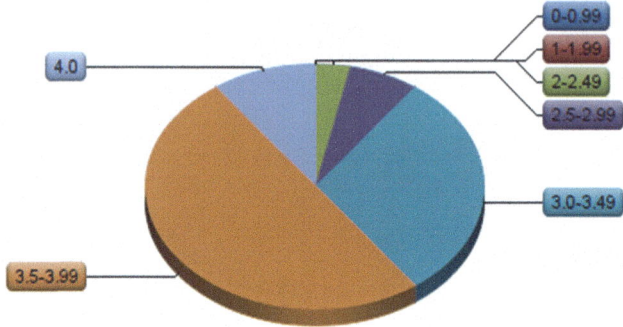

Fig. 9.7 Respondent GPA

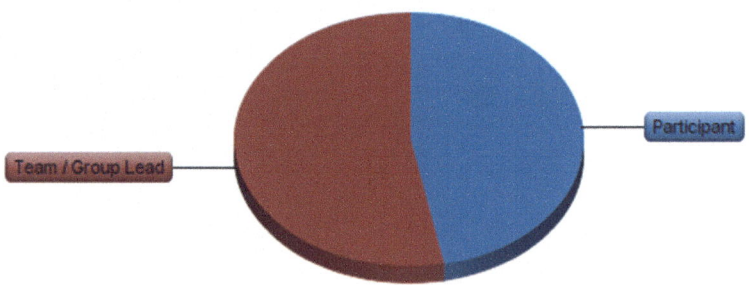

Fig. 9.8 Respondent participation type

10 hours per week on their participation and nearly 40 % indicated that they spent more than 20 hours per week on participation. The responses to this question are presented in Fig. 9.9.

Given the level of time commitment demonstrated by participants (in Fig. 9.9), an obvious question is why participants are willing to devote so much time to this activity (given that they could instead be devoting the time to other activities that may provide other types of professional or personal benefit). Table 9.1 presents an answer to this question: it details the participants' reasons for participating.

With the reasons why participants join a program now known, focus logically shifts to whether these benefits are being realized or not. An answer to this question is presented in Table 9.2, which indicates the areas where participants believed they have gained benefit through their participation.

A few dimensions of prospective benefit are now explored. Figure 9.10 asks participants whether they are interested in employment in their field of participation. Figure 9.11 indicates participants' responses to the question of whether they feel participation will aid them in seeking employment.

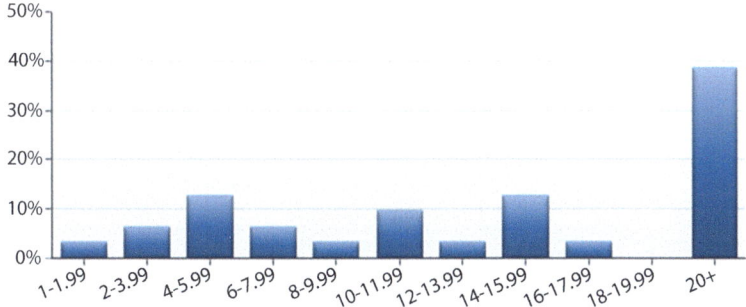

Fig. 9.9 Respondent weekly commitment

Table 9.1 Reasons for participating

Answer	%
Knowledge about spacecraft design	96
Knowledge about structured design processes	75
Knowledge about a particular technical topic	68
Knowledge about project management	82
Knowledge about time management	68
Leadership experience	79
Improving technical skills	93
Improving time management skills	64
Experience working with those from other disciplines	86
Real-world project experience	96
Item for resume	75
Improved presentation skills	61
Inclusion as author on technical paper	21
Experience working on a large group project	82
Experience with a structured design process	82
Experience related to a particular technical topic	61
Project management experience	64
Time management experience	57
Improving leadership skills	75
Improving project management skills	68
Understanding of how my discipline relates to others	64
Learn other discipline's technical details/terminology	68
Improved chance of being hired in desired field	75
Increased self-confidence	57
Ability to present at professional conference	43
Recognition in the university community	46

Table 9.2 Participant gains from participating

Answer	%
Knowledge about spacecraft design	96
Knowledge about structured design processes	92
Knowledge about a particular technical topic	88
Knowledge about project management	73
Knowledge about time management	73
Leadership experience	85
Improving technical skills	85
Improving time management skills	73
Experience working with those from other disciplines	88
Real-world project experience	92
Item for resume	85
Improved presentation skills	65
Inclusion as author on technical paper	27
Experience working on a large group project	85
Experience with a structured design process	77
Experience related to a particular technical topic	73
Project management experience	65
Time management experience	62
Improving leadership skills	81
Improving project management skills	73
Understanding of how my discipline relates to others	88
Learn other discipline's technical details/terminology	85
Improved chance of being hired in desired field	81
Increased self-confidence	77
Ability to present at professional conference	38
Recognition in the university community	65

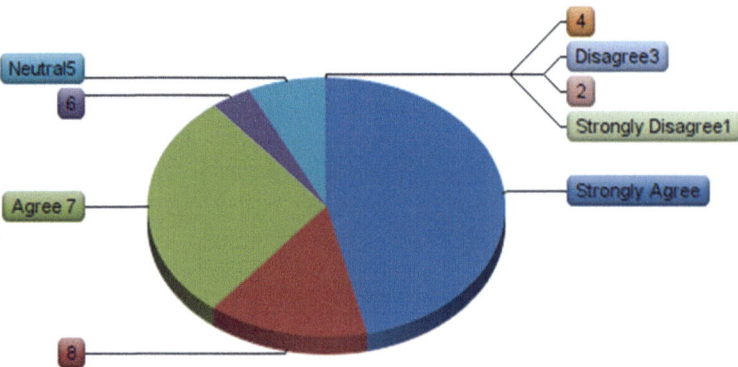

Fig. 9.10 Interest in employment in field of participation

9.6 Reporting This Value

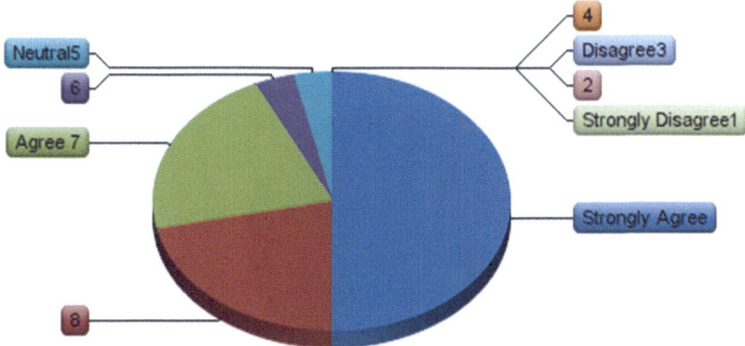

Fig. 9.11 Belief in participation aiding employment

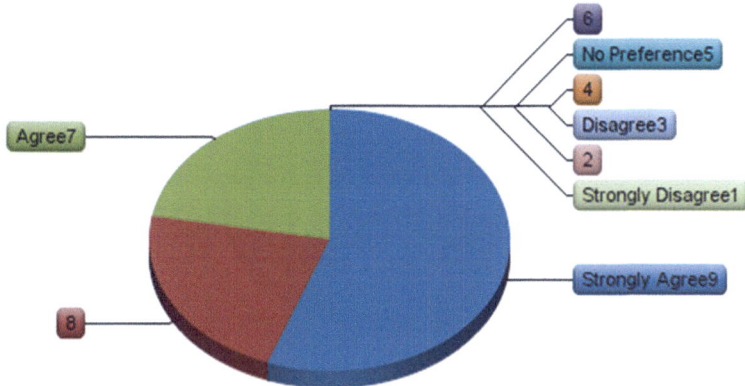

Fig. 9.12 Belief that participation has improved technical skills

Figures 9.12, 9.13, 9.14, and 9.15 ask participants to indicate whether they believe that participation has improved their skills and attitudes in various areas. Figure 9.12 asks respondents whether they believe that participation has aided their technical skills. Figure 9.13 asks respondents whether they feel that participation has increased their interest in space.

Figures 9.14 and 9.15 continue this line of inquiry. Figure 9.14 asks respondents whether they believe that participation has aided their project management skills and Fig. 9.15 asks about leadership skills.

Participants were then asked to discuss whether specific outcomes had occurred. These outcomes, which are based on the USRRA survey [101, 102], are presented in Table 9.3.

From the data presented, it is clear that small spacecraft programs deliver a wide variety of benefits to student participants. These range from the development of spacecraft-specific knowledge and skills to more general ones.

166   9 Summative Assessment

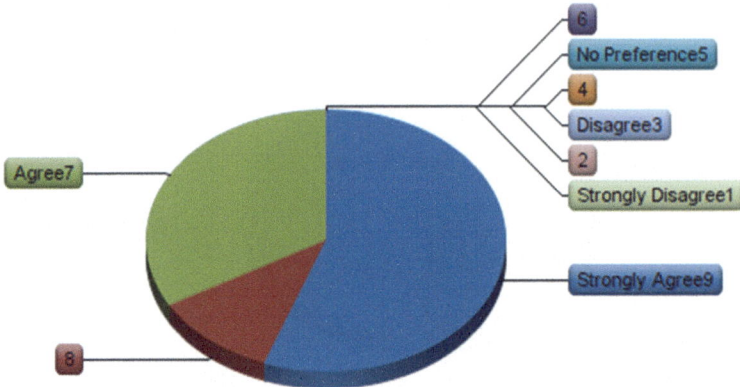

Fig. 9.13 Belief that participation has increased interest in space

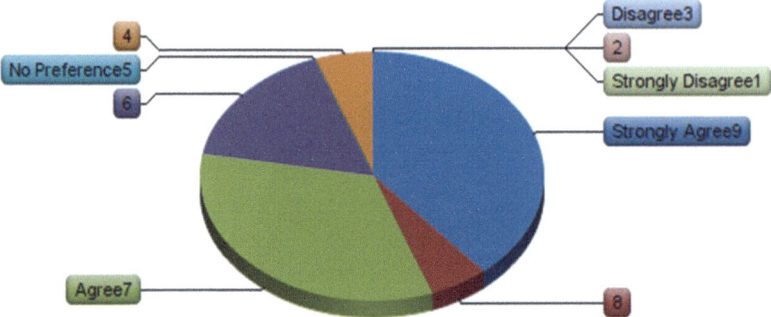

Fig. 9.14 Participation has increased project management skills

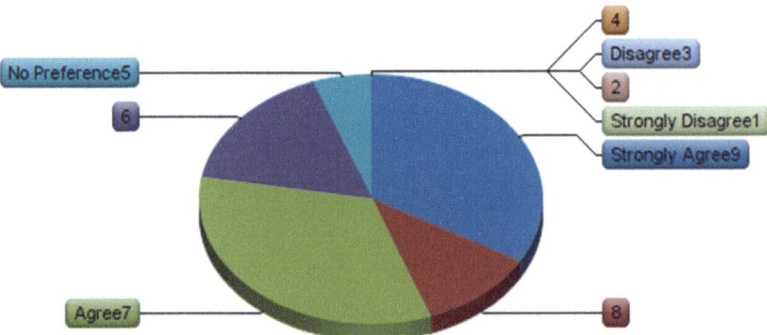

Fig. 9.15 Participation has increased leadership skills

Table 9.3 Outcomes

Answer		%
I presented a talk or poster to other students or faculty		100
I presented a talk or poster at a professional conference		57
I attended a conference		86
I wrote or cowrote a paper that was published in an academic journal		29
I wrote or cowrote a paper that was published in an undergraduate research journal		0
I will present a talk or poster to other students and faculty		43
I will present a talk or poster at a professional conference		29
I will write or cowrite a paper to be published in an academic journal		14
I will write or cowrite a paper to be published in an undergraduate research journal		0
I won an award or scholarship based on my research		14

9.7 Explaining Why the Program Is Important

Perhaps the most important part of summative assessment (whether anecdotal, qualitative, or quantitative) is to explain (and perhaps assess) the importance of a small spacecraft program. Programs may compete for funds, locally, nationally, or internationally, with other small spacecraft programs and with other prospective educational programs for students. Locally, they compete for resources ranging from laboratory and office space to volunteer-students time, as well. Given the comparative cost and duration of developing, testing, and launching a small spacecraft into space, explaining (and, better yet, proving) the particular benefits of each program will be essential to its success. They key to this is to determine the special elements of the program which differentiate it from other programs (small spacecraft or otherwise) and activities. These may fall neatly into a category such as engineering advances, scientific results, and educational value. In many cases, a more nuanced explanation may be necessary, as the program may deliver value in all three of these areas and assessment of the program in terms of a single area's metric may be inappropriate and fail to fully capture the value that the program provides.

9.8 Conclusion

This chapter has provided an overview of summative assessment. It has discussed the reasons for performing summative assessment as well as provided examples of the value that summative assessment can provide and examples of its use. The following chapter (Chap. 10) provides specific examples of prior summative assessment, including both examples of the mechanisms used and the results that have been generated.

References

1. Straub, J. 2013. OpenOrbiter: Analysis of a student-run space program. Presented at Proceedings of the 64th International Astronautical Congress.
2. Straub, J., S. Kerlin, and T. Stokke. 2014. How we're changing computer science education and how you can help. Presented at Proceedings of the 2014 Midwest Instructional Computing Symposium.
3. Straub, J., and D. Whalen. 2014. Evaluation of the educational impact of participation time in a small spacecraft development program. *Education Sciences* 4(1): 141–154.
4. Straub, J. 2015. Initial results from the first national survey of student outcomes from small satellite program participation. Presented at AIAA Space and Astronautics Forum and Exposition 2015.
5. Snell, K.D. 1996. The apprenticeship system in British history: The fragmentation of a cultural institution. *History of Education* 25(4): 303–321.
6. Elbaum, B. 1989. Why apprenticeship persisted in Britain but not in the United States. *Journal of Economic History* 49(2): 337–349.
7. Straub, J., J. Berk, A. Nervold, and D. Whalen. 2013. OpenOrbiter: An interdisciplinary, student run space program. *Advances in Education* 2: 4–10.
8. Mountrakis, G., and D. Triantakonstantis. 2012. Inquiry-based learning in remote sensing: A space balloon educational experiment. *Journal of Geography in Higher Education* 36(3): 385–401.
9. Mathers, N., A. Goktogen, J. Rankin, and M. Anderson. 2012. Robotic mission to mars: Hands-on, minds-on, web-based learning. *Acta Astronautica* 80: 124–131.
10. Nordlie, J., and R. Fevig. 2011. Blending research and teaching through high-altitude balloon projects. In *Proceedings of the 2nd Annual Academic High Altitude Conference*. Ames, IA.
11. Fevig, R., J. Casler, and J. Straub. 2012. Blending research and teaching through near-earth asteroid resource assessment. Presented at Space Resources Roundtable and Planetary & Terrestrial Mining Sciences Symposium.
12. Hall, S.R., I. Waitz, D.R. Brodeur, D.H. Soderholm, and R. Nasr. 2002. Adoption of active learning in a lecture-based engineering class. Presented at Proceedings of the 32nd Annual Frontiers in Education Conference.
13. Brodeur, D.R., P.W. Young, and K.B. Blair. 2002. Problem-based learning in aerospace engineering education. Presented at Proceedings of the 2002 American Society for Engineering Education Annual Conference and Exposition.
14. Ramsay, J., and E. Sorrell. 2007. Problem-based learning an adult-education-oriented training approach for SH&E practitioners. *Professional Safety* 52(9): 41–46.
15. von Kotze, A., and L. Cooper. 2000. Exploring the transformative potential of project-based learning in university adult education. *Studies in the Education of Adults* 32(2): 212–228.

16. Bütün, E. 2005. Teaching genetic algorithms in electrical engineering education: A problem-based learning approach. *International Journal of Electrical Engineering Education* 42(3): 223–233.
17. de-Camargo-Ribeiro, L.R. 2008. Electrical engineering students evaluate problem-based learning (PBL). *International Journal of Electrical Engineering Education* 45(2): 152–161.
18. Robson, N., I.S. Dalmis, and V. Trenev. 2012. Discovery learning in mechanical engineering design: Case-based learning or learning by exploring? Presented at 2012 ASEE Annual Conference.
19. Coller, B.D., and M.J. Scott. 2009. Effectiveness of using a video game to teach a course in mechanical engineering. *Computers & Education* 53(3): 900–912.
20. Das, S., S.A. Yost, and M. Krishnan. 2010. A 10-year mechatronics curriculum development initiative: Relevance, content, and results—Part I. *IEEE Transactions on Education* 53(2): 194–201.
21. Qidwai, U. 2011. Fun to learn: Project-based learning in robotics for computer engineers. *ACM Inroads* 2(1): 42–45.
22. Saunders-Smits, G.N., P. Roling, V. Brügemann, N. Timmer, and J. Melkert. 2012. Using the engineering design cycle to develop integrated project based learning in aerospace engineering. Presented at International Conference on Innovation, Practice and Research in Engineering Education.
23. Jayaram, S., L. Boyer, J. George, K. Ravindra, and K. Mitchell. 2010. Project-based introduction to aerospace engineering course: A model rocket. *Acta Astronautica* 66(9): 1525–1533.
24. Correll, N., R. Wing, and D. Coleman. 2013. A one-year introductory robotics curriculum for computer science upperclassmen. *IEEE Transactions on Education* 56(1): 54–60. doi:10.1109/TE.2012.2220774.
25. Broman, D., K. Sandahl, and M. Abu Baker. 2012. The company approach to software engineering project courses. *IEEE Transactions on Education* 55(4): 445–452. doi:10.1109/TE.2012.2187208.
26. Okudan, G.E., and S.E. Rzasa. 2006. A project-based approach to entrepreneurial leadership education. *Technovation* 26(2): 195–210.
27. Pollard, C.E. 2012. Lessons learned from client projects in an undergraduate project management course. *Journal of Information Systems Education* 23(3): 271–282.
28. Filcik, A., K. Bosch, S. Pederson, and N. Haugen. 2012. The effects of project-based learning (PBL) approach on the achievement and efficacy of high school mathematics students: A longitudinal study investigating the effects of the PBL approach in mathematics education. In *2012 NCUR*.
29. Milner-Bolotin, M., and M. Svinicki. 2012. Teaching physics of everyday life: Project-based instruction and a collaborative work in undergraduate physics course for nonscience majors. *Journal of the Scholarship of Teaching and Learning* 1(1): 25–40.
30. Doppelt, Y. 2003. Implementation and assessment of project-based learning in a flexible environment. *International Journal of Technology and Design Education* 13(3): 255–272.
31. Ayob, A., R.A. Majid, A. Hussain, and M.M. Mustaffa. 2012. Creativity enhancement through experiential learning. *Advances in Natural and Applied Science* 6(2): 94–99.
32. Simons, L., L. Fehr, N. Blank, H. Connell, D. Georganas, D. Fernandez, and V. Peterson. 2012. Lessons learned from experiential learning: What do students learn from a practicum/internship? *International Journal of Teaching and Learning in Higher Education* 24(3): 325–334.
33. Bauerle, T.L., and T.D. Park. 2012. Experiential learning enhances student knowledge retention in the plant sciences. *HortTechnology* 22(5): 715–718.
34. Edwards, A., S.M. Jones, E. Wapstra, and A.M. Richardson. 2012. Engaging students through authentic research experiences. Presented at Proceedings of the Australian Conference on Science and Mathematics Education (Formerly UniServe Science Conference).
35. Breiter, D., C. Cargill, and S. Fried-Kline. 2013. An industry view of experiential learning. *Hospitality Review* 13(1): 8.

36. Hotaling, N., B.B. Fasse, L.F. Bost, C.D. Hermann, and C.R. Forest. 2012. A quantitative analysis of the effects of a multidisciplinary engineering capstone design course. *Journal of Engineering Education* 101(4): 630–656.
37. Fasse, B., N. Hotaling, C. Forest, L. Bost, and C. Hermann. 2012. The case for multidisciplinary capstone design: A quantitative analysis of impact on job placement and product quality. Presented at Proceedings of the Biomedical Engineering Society (BMES) 2012 Annual Meeting. http://pbl.gatech.edu/wp-content/uploads/2013/05/fasse_bmes_2012.pdf.
38. Gilmore, M. 2013. Improvement of STEM education: Experiential learning is the key. *Modern Chemistry & Applications* 1: e109.
39. Larsen, J.A., J.F.D. Nielsen, and C. Zhou. 2013. Motivating students to develop satellites in problem and project-based learning (PBL) environment. *International Journal of Engineering Pedagogy* 3(3): 11–17.
40. Nielsen, J.F.D., X. Du, and A. Kolmos. 2010. Innovative application of a new PBL model to interdisciplinary and intercultural projects. *International Journal of Electrical Engineering Education* 47(2): 174–188.
41. Fortescue, P., G. Swinerd, and J. Stark. 2011. *Spacecraft Systems Engineering*, 4th ed. West Sussex: Wiley.
42. Wertz, J.R., D.F. Everett, and J.J. Puschell. 2011. *Space Mission Engineering: The New SMAD*. Hawthorne, CA: Microcosm Press.
43. Straub, J., and D. Whalen. 2013. An assessment of educational benefits from the OpenOrbiter space program. *Education Sciences* 3(3): 259–278.
44. Thakker, P., and G. Swenson. 2010. Management and implementation of a CubeSat interdisciplinary senior design course. In *Emergence of Pico- and Nanosatellites for Atmospheric Research and Technology Testing*, ed. P. Thakker and W. Shiroma. Reston: AIAA.
45. Swartwout, M. 2011. AC 2011-1151: Significance of student-built spacecraft design programs it's impact on spacecraft engineering education over the last ten years. Presented at Proceedings of the American Society for Engineering Education Annual Conference. http://www.asee.org/file_server/papers/attachment/file/0001/1307/paper-final.pdf.
46. ———. 2004. University-class satellites: From marginal utility to 'disruptive' research platforms. Presented at Proceedings of the 18th Annual AIAA/USU Conference on Small Satellites.
47. Deepak, R.A., and R.J. Twiggs. 2012. Thinking out of the box: Space science beyond the CubeSat. *Journal of Small Satellites* 1(1): 3–7.
48. Swartwout, M. 2013. Cheaper by the dozen: The avalanche of rideshares in the 21st century. Presented at 2013 IEEE Aerospace Conference.
49. ———. 2013. The long-threatened flood of university-class spacecraft (and CubeSats) has come: Analyzing the numbers. Presented at Proceedings of the 27th Annual AIAA/USU Conference on Small Satellites.
50. Bergsrud, C., and J. Straub. 2013. A 6-U commercial constellation for space solar power supply to other spacecraft. Presented at Spring 2013 CubeSat Workshop.
51. Padmanabhan, S., S. Brown, P. Kangaslahti, R. Cofield, D. Russell, R. Stachnik, J. Steinkraus, and B. Lim. 2013. A 6U CubeSat constellation for atmospheric temperature and humidity sounding. Proceedings of the AIAA/USU Conference on Small Satellites.
52. Bailey, J., S. Tsitas, D. Bayliss, and T. Bedding. 2012. A CubeSat mission for exoplanet transit detection and astroseismology. Presented at Proceedings of the 6U CubeSat Low Cost Space Missions Workshop.
53. Chirayath, V., and B. Mahlstedt. 2012. HiMARC 3D-high-speed, multispectral, adaptive resolution stereographic CubeSat imaging constellation. Presented at Proceedings of the AIAA/USU 2012 Small Satellite Conference.
54. Noca, M., F. Jordan, N. Steiner, T. Choueiri, F. George, G. Roethlisberger, N. Scheidegger, H. Peter-Contesse, M. Borgeaud, and R. Krpoun. 2009. Lessons learned from the first swiss pico-satellite: SwissCube. Lessons learned from the first swiss pico-satellite: SwissCube.

References

55. Weeks, D., A.B. Marley, and J. London III. 2009. SMDC-ONE: An army nanosatellite technology demonstration. Presented at SMDC-ONE: An Army Nanosatellite Technology Demonstration.
56. Abramowitz, L.R. Paul La Tour, Peter Mastro, Alan Frazier, Catherine Venturini, George Sondecker, and Lyle Abramowitz US air force's SMC/XR SENSE NanoSat program, , AIAA SPACE 2011 Conference & Exposition.
57. Taraba, M., C. Rayburn, A. Tsuda, and C. MacGillivray. 2009. Boeing's CubeSat TestBed 1 attitude determination design and on-orbit experience. Presented at Proceedings of the AIAA/USU Conference on Small Satellites.
58. Fitzsimmons, S., and A. Tsuda. 2013. Rapid development using Tyvak's open source software model.
59. Straub, J. 2012. Cubesats: A low-cost, very high-return space technology. In *Proceedings of the 2012 Reinventing Space Conference*. Los Angeles, CA.
60. Hunyadi, G., J. Ganley, A. Peffer, and M. Kumashiro. 2004. The university nanosat program: An adaptable, responsive and realistic capability demonstration vehicle. Presented at 2004 IEEE Aerospace Conference Proceedings.
61. Skrobot, G., and R. Coelho. 2012. ELaNa–Educational launch of nanosatellite: Providing routine RideShare opportunities. Presented at Proceedings of SmallSat Conference.
62. 13 February 2013. *Call for Proposals: Fly Your Satellite!* http://www.esa.int/Education/Call_for_Proposals_Fly_Your_Satellite.
63. Garvey, J., and E. Besnard. 2004. Development status of a nanosat launch vehicle. In *AIAA Paper (2004–4065)*.
64. Milliron, R. 2013. Interorbital's NEPTUNE dedicated SmallSat launcher: 2013 test milestones and launch manifest update. In *2013 Spring CubeSat Developers' Workshop*. San Luis Obispo, CA.
65. Berk, J., J. Straub, and D. Whalen. 2013. Open prototype for educational NanoSats: Fixing the other side of the small satellite cost equation. In *Proceedings of the 2013 IEEE Aerospace Conference*. Big Sky, MT.
66. Twiggs, R., and B. Malphrus. 2011. CubeSats. In *Space Mission Engineering: The New SMAD*, ed. J.R. Wertz, D.F. Everett, and J.J. Puschell, 803–821. Hawthorne, CA: Microcosm Press.
67. Muylaert, J., R. Reinhard, C. Asma, J. Buchlin, P. Rambaud, and M. Vetrano. 2009. QB50: An international network of 50 CubeSats for multi-point, in-situ measurements in the lower thermosphere and for re-entry research. Presented at ESA Atmospheric Science Conference. Barcelona, Spain.
68. London, J., M. Ray, D. Weeks, and B. Marley. 2011. The first US army satellite in fifty years: SMDC-ONE first flight results. Proceedings of the AIAA/USU Conference on Small Satellites.
69. Woellert, K., P. Ehrenfreund, A.J. Ricco, and H. Hertzfeld. 2011. CubeSats: Cost-effective science and technology platforms for emerging and developing nations. *Advances in Space Research* 47(4): 663–684.
70. California Polytechnic State University. 2009. CubeSat design specification, revision 12. California Polytechnic State University, San Luis Obispo, CA, 1 Aug 2009.
71. Hevner, R., W. Holemans, J. Puig-Suari, and R. Twiggs. 2011. An advanced standard for CubeSats. Proceedings of the AIAA/USU Conference on Small Satellites.
72. Korzh, A. 2013. What are we educating our youth for? *European Education* 45(1): 50–73.
73. Lagemann, E.C., and H. Lewis. 2012. *What is College for? The Public Purpose of Higher Education*. New York: Teachers College Press.
74. Tomlinson, M. 2012. Graduate employability: A review of conceptual and empirical themes. *Higher Education Policy* 25(4): 407–431.
75. O'Mahony, K., and T.N. Garavan. 2012. Implementing a quality management framework in a higher education organisation: A case study. *Quality Assurance in Education* 20(2): 184–200.

76. Brown, G.L. 2012. *Dissolving the iron triangle: Increasing access and quality at reduced cost in public higher education*. Masters Thesis, George Mason University. http://digilib.gmu.edu/dspace/bitstream/1920/7911/1/BrownG_thesis_2012.pdf, http://www.refworks.com/refworks2/?r=references|MainLayout::init#.
77. Baum, S., C. Kurose, and M. McPherson. 2013. An overview of American higher education. *The Future of Children* 23(1): 17–39.
78. Oreopoulos, P., and U. Petronijevic. 2013. Making college worth it: A review of the returns to higher education. *The Future of Children* 23(1): 41–65.
79. Vigdor, J. 2013. Solving America's math problem. *Education Next* 13(1): n1.
80. Hanushek, E.A., L. Woessmann, and P.E. Peterson. 2012. Is the US catching up? *Education Next* 12(4): n4.
81. Young, L., and T. Papinczak. 2012. Strategies for sustaining quality in PBL facilitation for large student cohorts. *Advances in Health Sciences Education* 21: 1–9.
82. Horn, M. 2012. Game changer. *Education Next* 12: 93–94.
83. ———. 2013. The transformational potential of flipped classrooms. *Education Next* 13: 78–79.
84. Alston, A.J., D. Cromartie, D. Wakefield, and C.W. English. 2009. The importance of employability skills as perceived by the employers of United States' land-grant college and university graduates. *Journal of Southern Agricultural Education Research* 59(1): 59–72.
85. Fuller, B., J. Wright, K. Gesicki, and E. Kang. 2007. Gauging growth: How to judge no child left behind? *Educational Researcher* 36(5): 268–278.
86. Straub, J., and C. Korvald. 2013. A review of online collaboration tools used by the UND OpenOrbiter program. Presented at Proceedings of the Collaborative Online Organizations Workshop at the Autonomous Agents and Multi-Agent Systems (AAMAS) 2013 Conference.
87. Torgerson, D., J. Straub, A. Mohammad, C. Korvald, and D. Limesand. 2013. An open-source scheduler for small satellites. Presented at SPIE Defense, Security, and Sensing.
88. Torgerson, D., C. Korvald, J. Straub, and J. Berk. 2013. The development of operating software for an OPEN small spacecraft. Presented at Midwest Instruction and Computing Symposium.
89. Torgerson, D., C. Korvald, J. Berk, J. Straub, and S. Kerlin. 2013. The development of control and scheduling software for a small spacecraft. *University of North Dakota Graduate School Scholarly Forum*.
90. Sand, J., K. Goehner, C. Korvald, J. Berk, and J. Straub. 2013. Payload processing aboard an open source software CubeSat. Presented at Spring 2013 CubeSat Workshop.
91. Brewer, J., B. Badders, J. Berk, and J. Straub. 2013. Work to-date on mechanical design for an open hardware spacecraft. Presented at Spring 2013 CubeSat Workshop.
92. Goehner, K., C. Korvald, J. Straub, and R. Marsh; University of North Dakota Graduate School Scholary Forum. 2013. The development of payload software for a small spacecraft.
93. ———. 2013. The development of payload software for a small spacecraft. Presented at Midwest Instruction and Computing Symposium.
94. Workman, E., N. Schmitz, C. Bergsrud, J. Berk, and J. Straub. 2013. Communications subsystem of the OpenOrbiter spacecraft. *University of North Dakota Graduate School Scholarly Forum*.
95. Trooien, K., A. Nervold, J. Berk, J. Straub, and S. Broedel. 2013. The role of communication in the student research project OpenOrbiter. *University of North Dakota Graduate School Scholarly Forum*.
96. Nervold, A., J. Straub, and J. Berk. 2013. OpenOrbiter: A student-run space program. *University of North Dakota Graduate School Scholarly Forum*.
97. Bryant, Z., M. Olson, C. Bergsrud, J. Berk, and J. Straub. 2013. A power generation system for the OpenOrbiter CubeSat-class spacecraft. *University of North Dakota Graduate School Scholarly Forum*.

98. Straub, J., C. Korvald, A. Nervold, A. Mohammad, N. Root, N. Long, and D. Torgerson. 2013. OpenOrbiter: A low-cost, educational prototype CubeSat mission architecture. *Machines* 1: 1–32.
99. Hung, W. 2011. Theory to reality: A few issues in implementing problem-based learning. *Educational Technology Research and Development* 59(4): 529–552.
100. Straub, J., and D. Whalen. 2013. Student expectations from participating in a small spacecraft development program. *Aerospace* 1(1): 18–30.
101. 2009. *Ethnography & Evaluation Research*. http://www.salgsite.org.
102. May 2009. *Undergraduate Research Student Self-Assessment (URSSA) FAQs*. http://www.colorado.edu/eer/downloads/URSSA_FAQs.pdf.
103. Straub, J., D. Whalen, and R. Marsh. 2014. Assessing the value of the OpenOrbiter program's research experience for undergraduates. *Sage Open* 2014.
104. Straub, J., and R. Marsh. 2014. Assessment of educational expectations, outcomes and benefits from small satellite program participation. Presented at 28th Annual AIAA/USU Conference on Small Satellites.

Chapter 10
Results of Prior Assessment Work and Its Utility

Knowledge of prior work may be helpful to justifying the creation of a program, provide a metric against which to compare other programs, and provide an example of assessment techniques. For these reasons, this chapter provides an overview of prior work on the assessment of small spacecraft programs' educational value.

The chapter begins with a discussion of formative evaluation and then presents a formative evaluation survey instrument to identify the reasons for students' participation. Then, the gain quantification survey (prospectively both a formative and summative tool) is discussed, in the context of formative evaluation. Next, summative evaluation is discussed and the utility of the gain quantification survey for summative assessment is considered. Following this, the Undergraduate Research Student Self-Assessment summative tool and meeting expectations survey instrument are presented. The general results of prior summative evaluation are then discussed, before concluding.

10.1 Discussion of the Use of Formative Evaluation

Formative assessment (see [4–7] for a full discussion) is an integral part of maximizing the value of educational activities for student participants. Using a formative assessment approach, one does not wait until the end of the educational experience to assess how well it is meeting its objectives (and, in some cases, to assess the correctness of the objectives themselves). Instead, assessment is performed at various points through the exercise to allow in-process data to shape the way that future parts of the exercise are carried out.

This chapter is based on, revises and extends the papers "Student Expectations from Participating in a Small Spacecraft Development Program" [1], *"Evaluation of the Educational Impact of Participation Time in a Small Spacecraft Development Program"* [2] *and "An Assessment of Educational Benefits from the OpenOrbiter Space Program"* [3].

In the context of the OOSDI small spacecraft development program, formative assessment was used to determine what goals student participants had and whether these goals and other pre-identified ones were being reached as the project progressed. Formative evaluation was also conducted using an assessment survey at the end of each semester which informed the planning of future semesters' activities.

The next sections present several of the tools used for formative evaluation. Section 10.2 presents the interest/reasons for participation survey and Sect. 10.3 presents the formative use of the dual-purpose (formative/summative) gain quantification survey.

10.2 Formative Evaluation Tools: Interest/Reason for Participation Survey

This section presents a survey that was used to assess prospective student participants' reasons for participation. The survey was given to those who were considering participating for the first time as well as to those returning for another semester who had participated previously. Participants were asked to indicate whether they had participated previously, allowing the interests of new participants to be juxtaposed with those of returning ones. The use of this survey facilitated the adaptation of a program on a semester-by-semester basis to meet the interests and needs of each new group of student participants. First, the design of the survey is presented and, then, the data that was collected from its use is presented and analyzed.

10.2.1 Experimental Design

The survey was administered to returning and prospective participants in the OOSDI program at the University of North Dakota. This survey, which was conducted anonymously, asked students for demographic information and then asked them to characterize their reasons for participating. These surveys were given at initial meetings used for recruiting new participants and at initial meetings of project groups.

The survey respondents included both undergraduate and graduate respondents. Respondents for the data set presented in this section included eighteen undergraduate and four graduate students. Of these respondents, four were in their first year of their academic program, eleven where second year students, and two were third year students. Three were in their fourth year of studies and two were in their fifth year. The undergraduates consisted of one freshman, five sophomores, seven juniors, and six seniors. Note that as a largely volunteer program, the demographic makeup of participants varies (sometimes significantly) from semester to semester. The current focus areas (e.g., this survey was performed at a time when a key focus was software development) also dictate the breakdown of the majors of students involved.

10.2 Formative Evaluation Tools: Interest/Reason for Participation Survey

Table 10.1 Correlation between number of semesters involved in the program and the five assessed metrics

Technical skill	Spacecraft design	Level of excitement	Presentation skills	Presentation comfort	Aggregate improvement
0.26	0.17	−0.13	0.24	0.21	0.22

The respondents to the survey were predominantly computer science majors (due to the aforementioned software focus). Table 10.1 shows the majors and minors of the participants (note that if an individual indicated multiple minors, they are counted in each category).

Students were also asked whether they had previously participated or not. Twelve individuals indicated previous participation, while eleven indicated that they had not participated previously. Note that the participation is nearly evenly split between returning participants and the newly joined. Past participants were asked to indicate the duration of their previous participation. Four students indicated participation for one semester, seven indicated two semesters of participation, two students indicated three semesters of participation, and two students indicated four semesters of participation. Whether students had previously or were planning to receive academic credit for their participation was also assessed. Eighteen students indicated that they had not participated previously for and were not participating for academic credit (and did not plan to do so). Three indicated participation or planned participation for a course project. One indicated participation or planned participation for an independent study project and one indicated participation or planned participation for other academic credit.

The breakdown of student participants is a function of various recruiting efforts pursued by those involved in the program. A strong recruiting effort, for example, to involve freshmen in the previous year may be largely responsible for the number of sophomores indicated and the high number of individuals with two semesters of previous participation (and probably affected the number indicating one semester as well). It is apparent that approximately one-half of those surveyed are returning participants and one-half are new participants. The number of opportunities for participating for academic credit has also expanded over time. In the first two semesters (under a thematically related precursor program), there were only two students who participated for academic credit; the third semester had three participants for academic credit and the forth semester has allowed seven individuals to participate for academic credit. This survey was taken prior to the establishment of one of the opportunities for for-credit participation, so several of the individuals who indicated that they were not participating for academic credit ended up doing so (some others had graduated and thus didn't take the survey). In addition, to participation as part of a senior design, junior design, or other whole-class participation opportunity, several students have had the opportunity to perform work on the project to satisfy a component requirement of a class. For example, three students created Architecture Analysis and Design Language (AADL) documentation for the project part of a software architecture course. This type of participation is not included in this category.

The first seven questions collected demographic data, while subsequent questions assessed student expectations from program participation. The responses to these questions are presented in the subsequent section.

10.2.2 Data Collected

In question eight, students were asked to select all of the benefits that they hoped to gain through their participation. The list of possible areas of benefit that could be selected is presented below. Students were also given the opportunity to write in other areas of benefit. This list is based on pre-identified project goals and other benefits that students indicated they believed they had received or would like to receive through other surveys [3] and anecdotally.

Knowledge about spacecraft design	Experience working on a large group project
Knowledge about structured design processes	Experience with a structured design process
Knowledge about a particular technical topic	Experience related to a particular technical topic
Knowledge about project management	Project management experience
Knowledge about time management	Time management experience
Leadership experience	Improving leadership skills
Improving technical skills	Improving project management skills
Improving time management skills	Understanding of how my discipline relates to others
Experience working with those from other	Learn other discipline's technical disciplines details/terminology
Real-world project experience	Improved chance of being hired in desired field
Item for resume	Ability to present at professional conference
Improved presentation skills	Ability to present at professional conference
Inclusion as author on technical paper	Recognition in the university community

The responses of students to this question are summarized in Figs. 10.1, 10.2 and 10.3. Figure 10.1 presents overall counts of the number of respondents who indicated that they hoped to gain each particular area of benefit. Figure 10.2 indicates the percentage of graduate and undergraduate students who indicated that they hoped to obtain each type of benefit. Figure 10.3 compares the responses of new entrants to those who have previously participated. Note that, in most cases, the expectations of previous participants and new entrants are closely correlated. This would tend to suggest that these expectations are being met (as existing participants would likely not believe that they would stand to gain benefits that they had not personally experienced or seen others experience during their previous participation).

In question nine, students were asked to rank their top three areas of benefit by importance. Figures 10.4 and 10.5 depict the responses to this question, with experience in a large group project, real-world project experience and improved technical skills ranking first through third.

10.2 Formative Evaluation Tools: Interest/Reason for Participation Survey

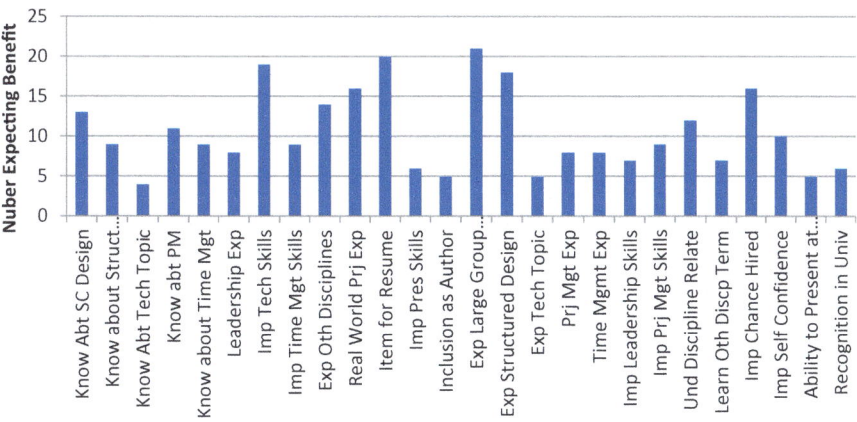

Fig. 10.1 Overview of benefits sought by participants

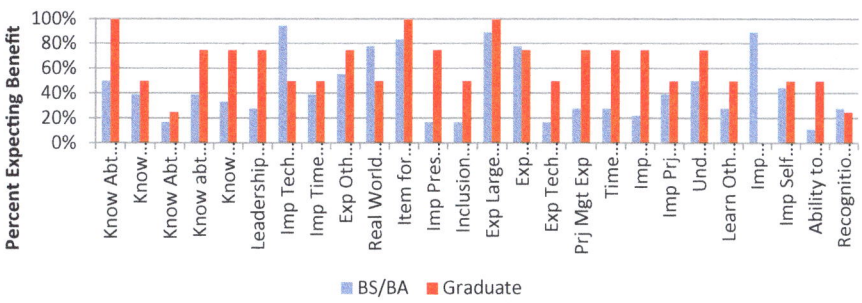

Fig. 10.2 Benefits sought by participants, by undergraduate/graduate status

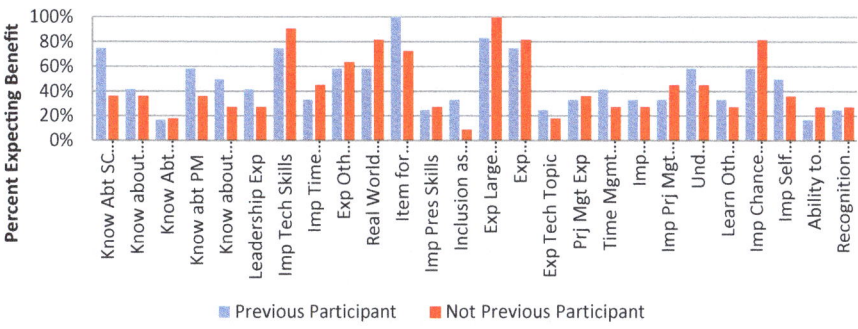

Fig. 10.3 Benefits sought by participants, by whether they have previously participated

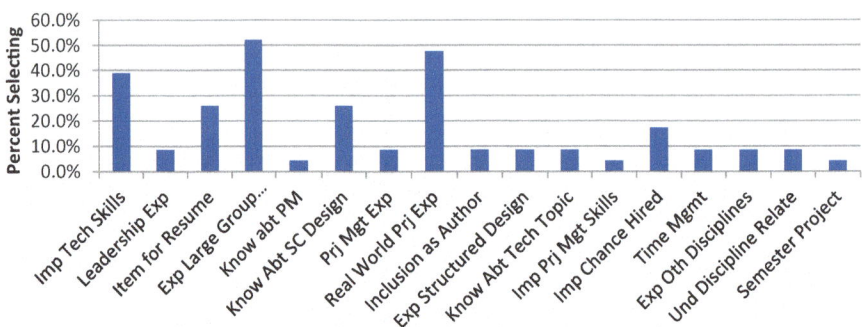

Fig. 10.4 Choices selected by respondents as one of their top three areas of desired benefit

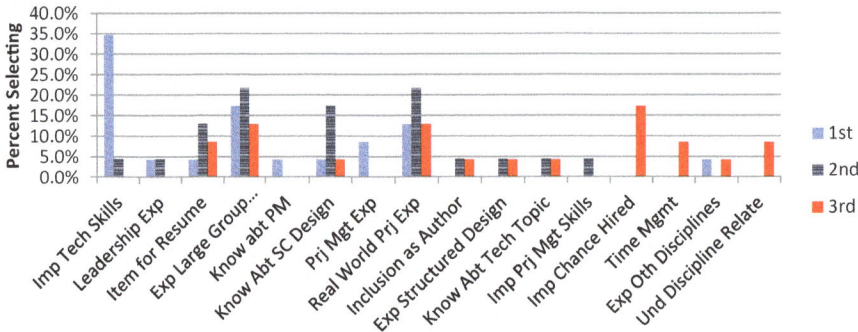

Fig. 10.5 Choices selected by respondents for each of the top three areas of desired benefit

The next four questions sought to assess specific reasons for students joining. Questions ten and eleven asked students whether they were interested in seeking employment in the field that they were or planned to participate in and whether they believed that participation would aid them in securing employment. Both of these questions solicited responses using a nine-point scale, ranging from 9-Strongly Agree to 5-Neutral to 1-Strongly Disagree. In both cases, the favorable answer (interest in seeking employment and participation aiding in securing employment) would correlate with the 9–5 scale range, while those believing the opposite would indicate between 5 and 1. Figures 10.6 and 10.7 present a histogram of the responses to these questions. The responses of undergraduate and graduate students are compared. Note that the responses of the graduate student respondents are generally less favorable than those of the undergraduates. This may be attributable to different career aspirations, existing experience levels (and thus less perception of additional benefit to be gained), or other factors. Exploration of the reason for these responses is a prospective subject for future work.

Finally, respondents were asked to indicate specific factors that drove their decision to join (in question 12). The responses to this question are presented in Fig. 10.8.

10.2 Formative Evaluation Tools: Interest/Reason for Participation Survey   181

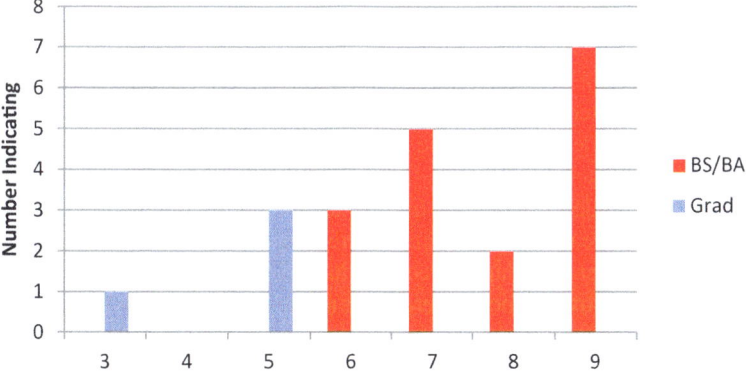

Fig. 10.6 Participants' response regarding whether they are seeking employment in this field

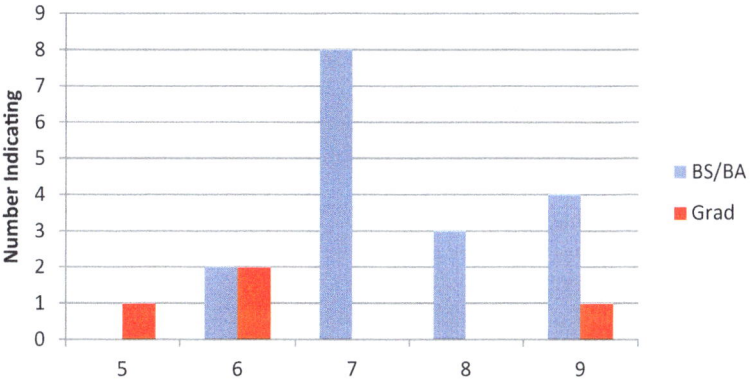

Fig. 10.7 Participants' response regarding whether they believe participating will aid employment

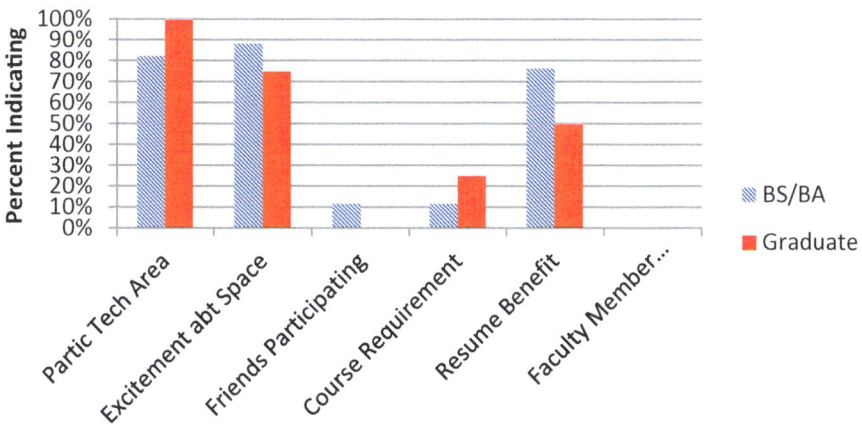

Fig. 10.8 Reason for participating, by undergraduate/graduate status

10.2.3 Analysis of Data

The data presented in the foregoing section indicates that students seek to attain a wide variety of related benefits from their participation in a small spacecraft program. Figure 10.1 demonstrates the breadth of these varied interests. The raw response data indicates that this is indicative of both the breadth of interest of individuals (with 12 students indicating over 10 areas of desired benefit and 4 of these indicating over 20 areas) and the diverse interests of the group as a whole. The areas of desired benefit were similar between graduate and undergraduate students, with graduate students showing more interest in most areas (some significantly, such as spacecraft design where all of the graduate students indicated an interest in the learning benefit as compared to only half of undergraduates). Undergraduates, however, showed significantly more interest in improved technical skills, real-world project experience, and improving their chances of getting hired (with nearly 90% of undergraduates indicating that they desired this benefit as compared to no graduate students). They also showed a marginally higher level of interest in experience in a structured design process and university recognition.

The expectations of those who had previously participated versus those who had not were close in most areas. Previous participants showed a significantly greater interest in gaining knowledge about spacecraft design and project management as well as a desire to build their resume. New participants showed a significantly greater interest in improving their technical skills, large project experience, and chances of being hired. The difference with regards to the last two may be due to the fact that participation in the project (at all) offers much of this type of benefit in a short period of time, meaning that there is less expectation of gaining it for existing participants.

Figures 10.4 and 10.5 demonstrate the relative importance of a few key areas of focus. The most important would appear to be improving technical skills. While more individuals selected large group project and real-world project experience, improving technical skills was selected by nearly 35% of respondents as the most important item (less than 20% of respondents selected large group project experience as their most important choice and less than 15% selected real-world project experience as most important). Knowledge about spacecraft design and resume-related benefits both were a top choice of over 20% of respondents (though these were divided over the three categories, with both having the highest interest shown in the second most important slot).

Two-thirds of respondents (66.7%) indicated an agreement or more favorable response (strongly agree was the mode of the responses to the question), indicating that they are seeking employment in a field related to their participation. An additional 14% indicated a less certain positive (6) response, for over 80% of respondents indicating some level of interest in participation-related employment. Of the remaining 20%, 14% indicated a neutral response and one individual indicated a disagree (3) response. Notably, the stronger interest responses came from undergraduates. There was a similarly strong response with regards to belief that participation would aid employment with no respondents indicating disagreement and

76 % indicating an agreement or stronger level response (19 % indicated a less positive agreement and one individual indicated a neutral response).

Finally, the responses to the reason for participating indicated strong correlation between undergraduate and graduate respondents. Over 80 % of undergraduates and all graduate students indicated that participation was based on interest in a particular technical area. Eighty-eight percent of undergraduates and 75 % of graduate students indicated participation due to space excitement. Twelve percent of undergraduates and no graduate students indicated that they were participating due to the fact that a friend was participating. Just over 10 % and 20 % of undergraduates and graduates, respectively, indicated they were participating to satisfy a course requirement. Seventy-six percent of undergraduates and 50 % of graduate students indicated that they joined to attain a resume benefit. No students from either group indicated that they were attracted by the participation of a particular faculty member. The foregoing shows two clear areas of focus and the tertiary focus area of the resume benefit, with the other areas being of less importance.

It is important to note that the limited number of respondents and the fact that they are all participants/prospective participants in a single small spacecraft program limits the potential for extrapolation from this data. The results can be compared to the national survey results discussed in Chap. 9 and presented in [8].

10.2.4 Summary

This section has presented an analysis of the reasons why students participate in small spacecraft development, based on surveys of prior and new participants in the University of North Dakota's OOSDI. It has demonstrated that students seek specific benefits from their participation and suggested that these benefits are being delivered, due to the correlation between the expectations of prior and new participants.

10.3 Formative Evaluation Tools: Gain Quantification Survey

The next formative evaluation tool that will be presented is the use of the gain quantification survey (a dual-mode formative/summative instrument) for formative use. This survey assesses both how students believe their skills and abilities have improved and their attribution of these enhancements to the program.

10.3.1 Benefits of Interdisciplinary Projects

A key area of assessment is the interdisciplinary nature of small spacecraft projects. Interdisciplinary projects are a typical feature of the modern workplace. Most undertakings of any size cannot be performed exclusively by practitioners of a single specialty. However, most student projects in an academic environment are performed within the context of a course or a degree program. Because of this, they generally involve a set of similarly trained students working on a narrowly defined topic. Even projects that span disciplines (e.g., teams participating in NASA's Lunabotics competition [9]) may be limited to only closely related disciplines (e.g., electrical, mechanical, and computer engineering).

Because of this, students may not gain exposure to a true interdisciplinary project (characterized by multiple specialists collaboratively performing work related to their area of specialty) until after they enter the workforce. This may require them to unlearn practices and approaches learned while working only in discipline-constrained teams. They may also experience frustration if the process of getting up-to-speed in this impairs their performance during their initial period (normally including some sort of an evaluation/probation process) with a new employer whom they seek to impress.

Involving students in interdisciplinary work prevents 'silo'-type work habits from developing; students instead learn how to work well in collaboration with others with skills divergent from their own. In addition to these general benefits, students also begin to learn the particular vernacular and work styles of the disciplines whose practitioners-in-training they collaborate with. Interdisciplinary projects may also be able to have a larger scale than those within a single discipline, offering an opportunity for project management practices and discipline-specific multiperson collaboration techniques (e.g., software version control management) to be learned and refined. All of this increases student participant preparation for workplace entry and success. These interdisciplinary benefits are a key area of focus and thus of assessment using the gain quantification survey and other tools.

10.3.2 Learning Objectives

Five main objectives were identified prior to beginning the OOSDI program. These were increasing proficiency in area-specific technical skills, spacecraft design and development skills, and presentation skills. The program also sought to increase excitement about space and participant comfort giving presentations. Each is now discussed.

Participants gaining area-specific technical skills is an obvious outcome from the spacecraft development program. For many students, the skills that have been and will (from future involvement) be enhanced are aligned with their major (or perhaps minor). Some students, however, opted to participate in an area different from their academic work to gain an understanding of and experience in a different field. The skills gained or enhanced through program participation were, of course, different for each group and, possibly, each individual (based on what tasks they worked on).

Learning spacecraft design and development skills was another obvious outcome of the program, due to the program focus on small spacecraft design and development. For many students, this was their first exposure to this topic. The skills imparted included iterative spacecraft design and refinement and subsystem-specific design and development skills. Perhaps the single largest lesson taught was with regards to the constraints that the space environment and launch and other costs place on the mass and volume (and, as a consequence of this, virtually all aspects) of the spacecraft. This topic also incorporated key interdisciplinary themes.

Presentation skills and comfort giving presentations were identified as key preparations for workforce success that could be enhanced by participation in this program. Success in the workplace environment requires effective use of written and verbal communications to convey highly technical information and other details. Of course, these skills cannot provide value if the individual doesn't put them to use. Given this, skill development and creating comfort using these skills were identified as key things that could be enhanced through program participation.

Enthusing participants about space and space engineering was also identified as a desired outcome. This outcome cannot be directly traced to future workplace success requirements. However, it was a necessity for project success, as this excitement was seen as a key driver for participants to remain involved in the project. Prospective future funding sources (e.g., NASA) for a national expansion of this type of work also define this is an evaluative criterion for proposal selection making delivering on this goal highly desirable for this purpose as well.

10.3.3 Data and Analysis

To assess the performance of OOSDI in attaining these educational objectives, a survey instrument was designed and administered to program participants across all of the groups at regularly scheduled meetings. For the data set reported here, 20 individuals completed the survey. These individuals included students studying computer science, electrical engineering, entrepreneurship, and space studies. This is a subset of the overall participation in the project which varied with nearly 300 students attending at least one meeting, and a smaller number (which fluctuated during the period described with, generally, between 45 and 75 students attending weekly group and/or general meetings). These results are now presented.

10.3.4 Overall Results

The initial survey asked participants to evaluate their status prior to project participation and at present for each of the five key outcome areas (technical skill, spacecraft design comfort, excitement about space, presentation skills, and presentation comfort). Participants were asked to respond on a nine-point scale for all status questions. Questions were given in the format:

On a scale of 1–9, _____ before starting work on the project:

On a scale of 1–9, _____ at the present time:

For each question, the above blanks were filled in with the particular item of focus. For example, for questions 13 and 18 the phrase "please rate your technical skill in your area of focus" was filled in resulting in the questions "on a scale of 1–9, please rate your technical skill in your area of focus before starting work on the project" and "on a scale of 1–9, please rate your technical skill in your area of focus at the present time." For this question, response choices ranged from 9-expert to 5-average to 1-novice. This scale was also used for questions 16 and 21 ("on a scale of 1–9, please rate your level of presentation skills").

For questions 14 and 19 ("on a scale of 1–9, please rate your level of comfort with spacecraft design"), response choices ranged from 9-very comfortable to 5-somewhat comfortable to 1-not comfortable. This scale was also used for questions 17 and 22 ("on a scale of 1–9, please rate your level of comfort with giving a presentation").

For questions 15 and 20 ("on a scale of 1–9, please rate your level of excitement with space before starting work on the project"), response choices ranged from 9-very excited to 5-average to 1-novice.

The average responses for each category, before and after participation, are presented in Fig. 10.9a. The average improvement by category is presented in Fig. 10.9b. There were a few isolated cases where participants reported lower status levels after participation as compared to before. For the skill questions, this type of response made no practical sense as there was no conceivable way that the project could have caused someone to regress in his/her skill level. On the excitement about space and comfort presenting questions, it is of course possible that these attitudes have declined during the time (due to program participation or otherwise). In each instance, the corresponding program impact question showed an average response (4–6 range) so it is presumed that these may be indicative of a change not caused by the program or perhaps participants not correlating their two responses.

Clearly, it is unrealistic to expect participants to improve in every category; some individuals may have had no or less involvement with areas of the project relevant to a particular category (e.g., presentations). It is thus also useful to look at how much skills improved for individuals who showed some improvement. Figure 10.10a

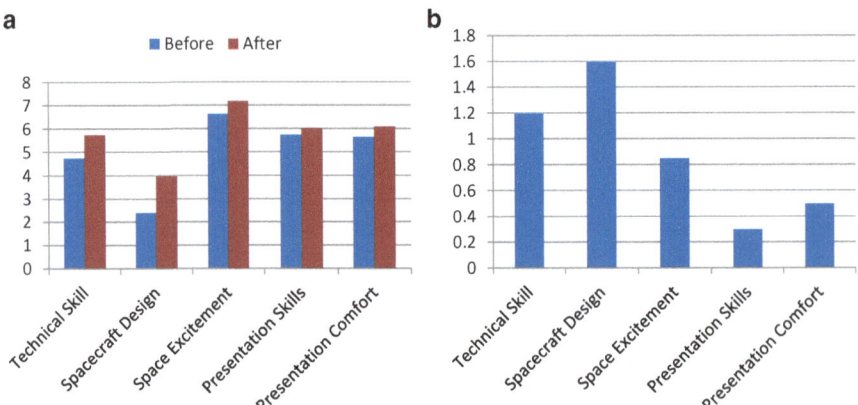

Fig. 10.9 (a) Comparison of beginning and ending status levels. (b) Improvement by status, average

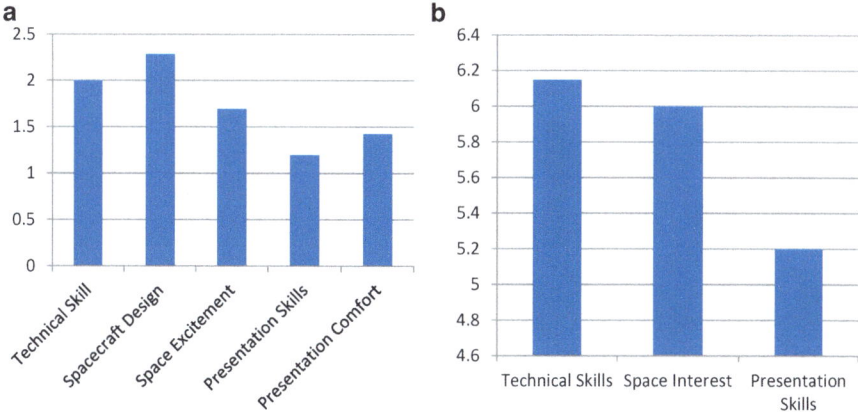

Fig. 10.10 (**a**) Average improvement by status for students showing improvement. (**b**) Attribution of program effect on creating change in status level

presents the average improvement for individuals showing improvement in each category.

In addition to asking respondents to characterize their pre-participation and post-participation skill levels, they were also asked to characterize the impact of the program on effecting this change. Again a nine-point scale was used with responses ranging from 9-strongly agree to 7-agree to 5-no preference to 3-disagree to 1-strongly disagree. Each of the three questions (23–25) was presented in the format:

Participation in this project has improved my _____:

Question 23 asked about "technical skills." Question 24 had respondents characterize the projects' impact on their "interest in space." Question 25 asked about "presentation skills."

The average responses to these questions are presented in Fig. 10.10b. Note that in all cases, the average is on the agree side, to varying extents. One individual who indicated that he or she hadn't "really done much" with regards to the project in the open ended question (number 26) influenced this somewhat, with this person's response excluded the response rise from 6.15 to 6.32, 6 to 6.16, and 5.2 to 5.32, for the technical skills, space interest, and presentation skills.

10.3.5 Comparison of Results Between Undergraduate and Graduate Students

As part of the survey instrument, participants were asked a variety of questions relevant to characterizing their academic status and involvement with the project. The next several subsections look at starting and ending status levels and the project's impact in terms of these conditions. This section characterizes these items by whether students were undergraduates or graduate students.

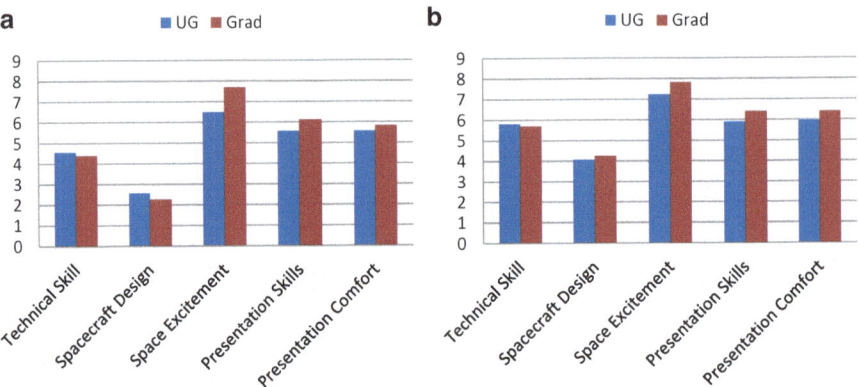

Fig. 10.11 (**a**) Beginning status levels, compared between graduate and undergraduate students. (**b**) Ending status levels, compared between graduate and undergraduate students

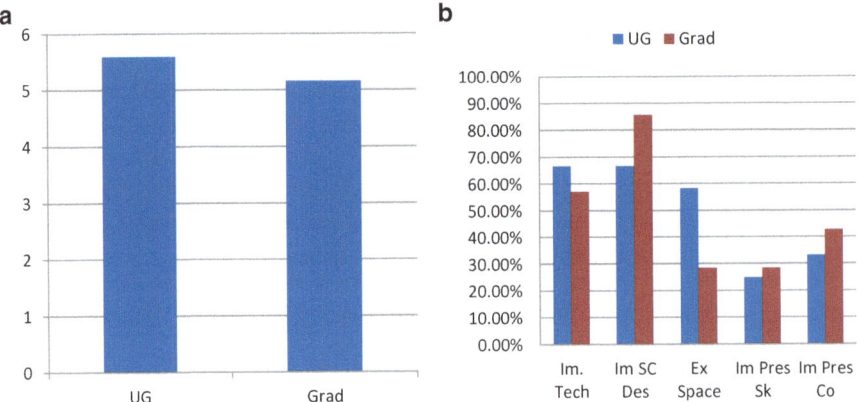

Fig. 10.12 (**a**) Average aggregate improvement, compared between graduate and undergraduate students. (**b**) Percentage of participants showing improvement in each status, compared between graduate and undergraduate students

Figure 10.11a presents the pre-participation levels for each category. Figure 10.11b presents these levels after participation. As these figures demonstrate, the relative levels of pre and post status are fairly consistent between undergraduates and graduate students. Graduate students average higher status levels for space excitement, presentation skills, and presentation comfort, prior to participation (undergraduates start marginally higher in the other categories). In spacecraft design, graduate students overtake undergraduates during participation. In all other cases, the group that started with a higher skill level also ended with a higher skill level.

Figure 10.12a depicts the relative average aggregate improvement (the average of the sum of the improvement values reported by each individual) between the two

10.3 Formative Evaluation Tools: Gain Quantification Survey 189

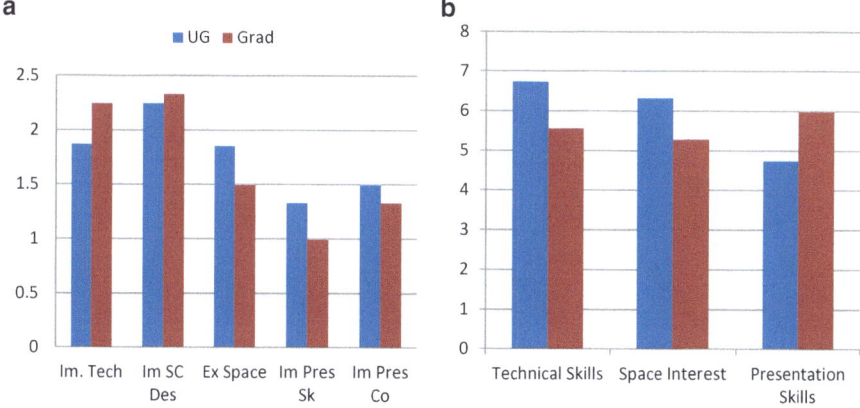

Fig. 10.13 (a) Average improvement in status levels (for students showing improvement), compared between graduate and undergraduate students. (b) Effect of program on causing improvement by status, compared between graduate and undergraduate students

groups. The one previously mentioned individual who reported he or she hadn't "done much with this project" was included in the graduate students. Excluding this individual raises the average to 6.2 (from 5.17) for the graduate students, which significantly exceeds the level reported by the undergraduates. Note that the individuals whose sum was a negative (decline) score have been excluded from this average. The negative value was excluded in the case of individuals who had other positive scores. Figure 10.12b shows the percentage of individuals in each category that had an improvement in each particular area.

In Fig. 10.13a, the average improvement for each status has been depicted for both graduate students and undergraduates. Figure 10.13b shows the responses related to program impact. In two of the three instances (technical skill and space interest) more improvement is shown for undergraduates as compared to graduate students. In the third, presentation skills, significantly more improvement is shown for graduate students. Excluding the individual who indicated a lack of participation, technical skills rise from 5.57 to 6 (as compared to 6.75 for undergraduates), space interest rise from 5.28 to 5.67 (as compared to 6.33 for undergraduates), and presentation skills rise from 6 to 6.5 (as compared to 4.75 for undergraduates).

10.3.6 Comparison of Results Between Team Leads and Participants

The relative performance impact of the project on individuals who are team leads versus on those who are not is now considered. Figure 10.14a presents the pre-participation status levels for both team leads and non-team leads. Figure 10.14b presents the post-participation status levels.

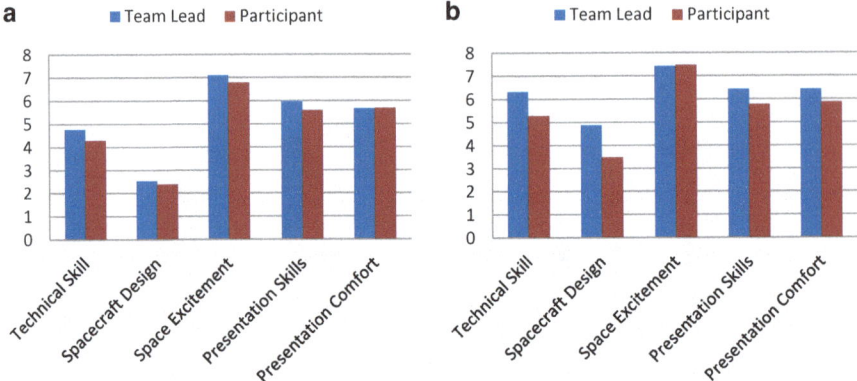

Fig. 10.14 (**a**) Beginning status levels, compared between team leads and participants. (**b**) Ending status levels, compared between team leads and participants

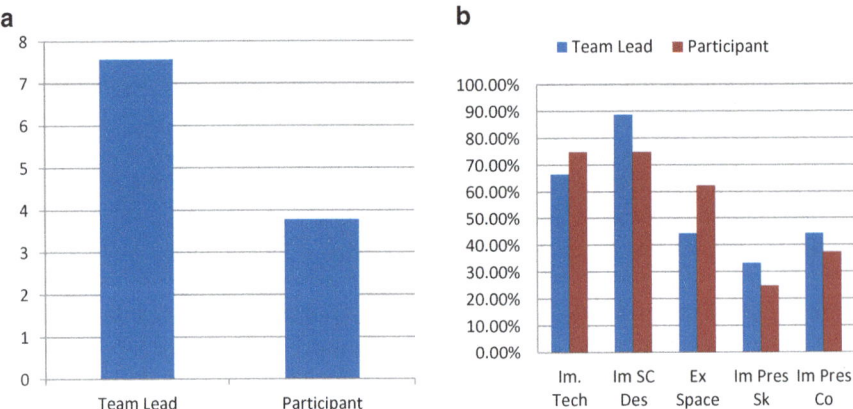

Fig. 10.15 (**a**) Average aggregate improvement, compared between team leads and participants. (**b**) Percentage of participants showing improvement in each status, compared between team leads and participants

The average aggregate improvement for team leads versus non-lead participation is depicted in Fig. 10.15a. This shows that team leads enjoyed over double the benefit of participation as compared with non-lead participants (7.57 vs. 3.78). Excluding the one individual who indicated a lack of involvement increases the participant average to 4.25. Figure 10.15b depicts the percentage of participants showing improvement in each category for both team leads and non-lead participants. A higher percentage of leads showed improvement in spacecraft design, presentation skills, and presentation confidence. A higher percentage of non-leads showed improvement in technical skills and excitement about space.

In Fig. 10.16a, b the level of improvement for each category and the effect of the program are considered. The average improvement shown by the team leads exceeds

10.3 Formative Evaluation Tools: Gain Quantification Survey

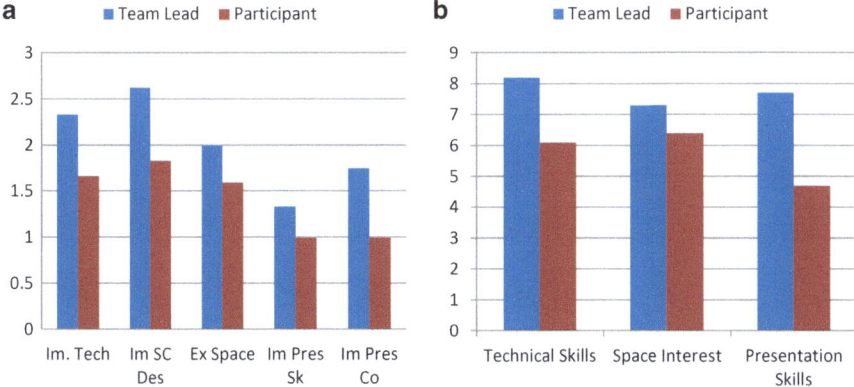

Fig. 10.16 (a) Improvement in status levels for those showing improvement in each category, compared between team leads and participants. (b) Effect of program on causing improvement by status, compared between team leads and participants

the level shown by the non-lead participants across all categories. The attribution of the program causing improvement is also higher across all categories is also higher for the team leads.

The data presented clearly indicates that team leads enjoyed significantly more benefit from participation as compared to the non-lead participants. Not only did they show significantly greater benefit (slightly over double), but they attributed this benefit to participation in the program to a greater extent.

10.3.7 Comparison of Results by Level of Weekly Participation

The impact of how much time is spent per week on the project is now considered. Respondents were asked to characterize their participation on the project into one of three categories: 1–3.99 hours per week spent, 4–7.99 hours per week spent, or 8+ hours per week spent. Figure 10.17a, b shows the pre-participation and post-participation status levels.

The average aggregate improvement, by level of weekly participation, is depicted in Fig. 10.18a. A correlation between greater work on the project and improvement is shown with those working 1–3.99 h showing an average aggregate improvement of 4.36 (4.8 with the individual who indicated minimal participation excluded) as compared to 7.75 for those spending 4–7.99 h and 8 for those spending 8 or more hours per week on the project. Figure 10.18b indicates the percentage of participants showing improvement for each category in each condition.

In Fig. 10.19a the average level of improvement in each category is depicted. Greater improvement is seen in all categories for the 4–7.99 as opposed to the 1–3.99 category. Due to the limited number of individuals responding in the 8+ category, the improvement is centered in two categories (with most of the improve-

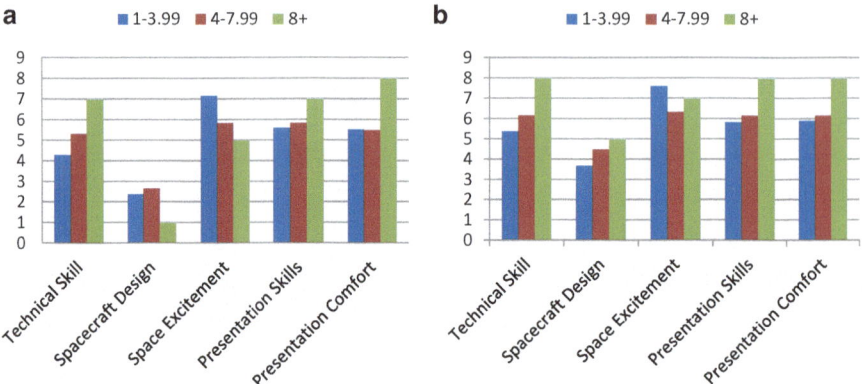

Fig. 10.17 (**a**) Beginning status levels, compared between weekly levels of participation. (**b**) Ending status levels, compared between weekly levels of participation

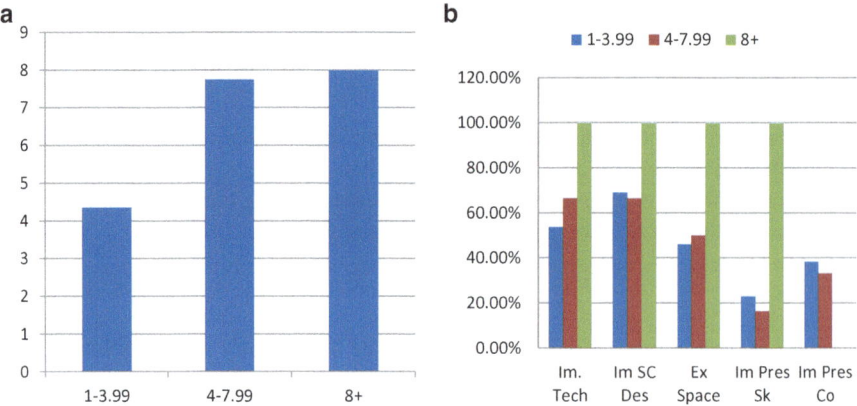

Fig. 10.18 (**a**) Average aggregate improvement, compared between weekly levels of participation. (**b**) Percentage of participants showing improvement in each status, compared between weekly levels of participation

ment being located in spacecraft design). Other categories underperform the 3–7.99 and 1–3.99 groups.

The impact of the program on causing the indicated improvement is now considered. The 8 hours per week or more category shows greater attribution of results to the program in each category (as compared to the 1–3.99 and 4–7.99 conditions). The 4–7.99 condition shows more attribution (as compared to the 1–3.99 condition) in two categories (technical and presentation skills), while the 1–3.99 condition shows greater attribution in the space interest category.

The foregoing shows a clear correlation between the amount of time spent weekly on the project and improvement. This is most pronounced between the 1–3.99 and 4–7.99 conditions with only minimal (average) improvement being seen between the 4–7.99 and 8+ categories.

10.3 Formative Evaluation Tools: Gain Quantification Survey

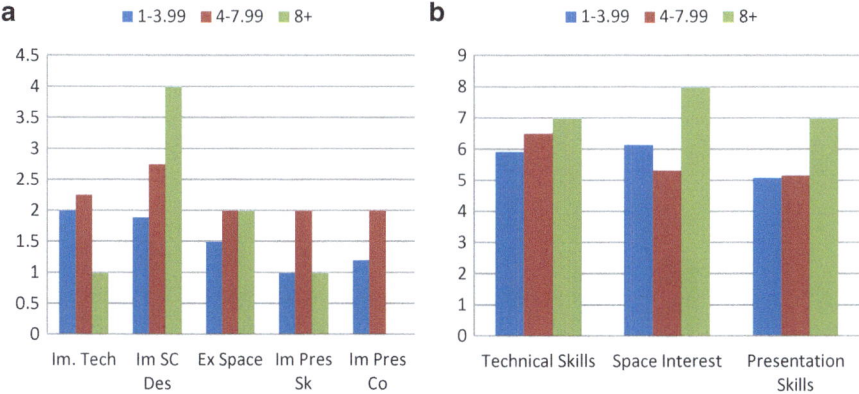

Fig. 10.19 (**a**) Improvement in status levels, compared between weekly levels of participation. (**b**) Effect of program on causing improvement by status, compared between weekly levels of participation

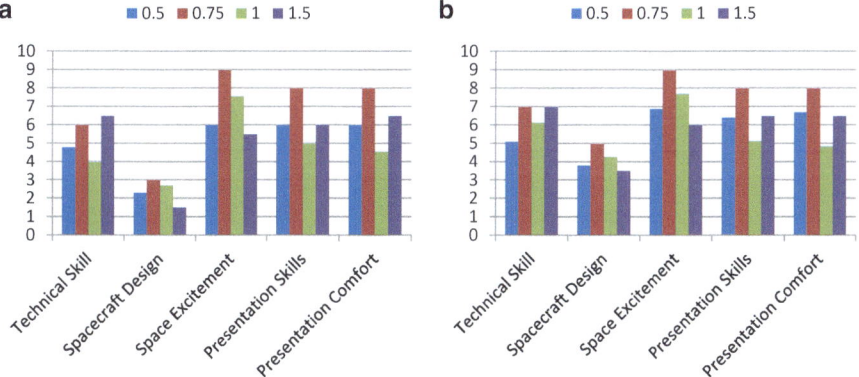

Fig. 10.20 (**a**) Beginning status levels, compared by time participating (in academic years). (**b**) Ending status levels, compared by time participating (in academic years)

10.3.8 Comparison of Results by Amount of Time Participating

Correlations between the duration of participation (how long it has been since the individual commenced participation) and results are now assessed. Figure 10.20a, b shows the pre-participation and post-participation status values. There is little time-category correlation demonstrated, as would be expected.

Figure 10.21a shows the correlation between the duration of participation and average aggregate improvement. A marginal increase is seen between 0.5 years and 1 year. One individual indicated 0.75 years of participation (via writing this answer in on the survey sheet; it was not a choice) and showed a comparative underperfor-

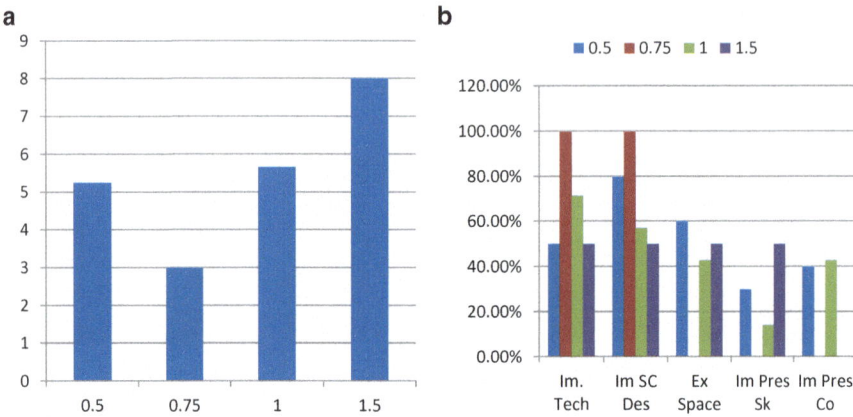

Fig. 10.21 (a) Average aggregate improvement, compared by time participating (in academic years). (b) Percentage of participants showing improvement in each status, compared by time participating (in academic years)

mance (relative to the 0.5- and 1-year categories). Excluding the individual who indicated minimal involvement, the 0.5 year average increases to 6 (from 5.25), surpassing the 5.66 response from the 1-year category. The 8 average increase from the 1.5-year condition still surpasses both.

It would seem that there is some correlation between the time spent involved and the average level of increase; however, this cannot be stated definitively for a number of reasons. First, it appears there was some confusion related to responses in this category altogether due to the ambiguity between calendar years and academic years. Second, the inclusion and exclusion of outlier, erroneous, and ambiguous data points appears to have a particular effect in this category with the exclusion of the individual indicating limited involvement bringing the average of the 0.5-year participants above that of the 1-year participants. Another data point (where the individual indicated agreement/agreement-strong agreement with the statements regarding impact but didn't indicate skill improvement), if excluded, would raise the 1-year condition to 6.8, bringing the two back into stronger correlation.

Figure 10.21b shows the correlation between the amount of time participating and the percentage of individuals showing improvement in each category. The limited membership of several categories makes this graph very erratic. Figure 10.22a, b shows the improvement in status levels, by category and attribution by category for each duration of participation condition. Again, the limited membership of some conditions makes both of these graphs somewhat erratic.

It would appear that there is a correlation between the duration that the participant has been involved and the level of benefit attained. However, possible ambiguity in the question and limited membership in certain conditions have made this conclusion uncertain.

10.3 Formative Evaluation Tools: Gain Quantification Survey   195

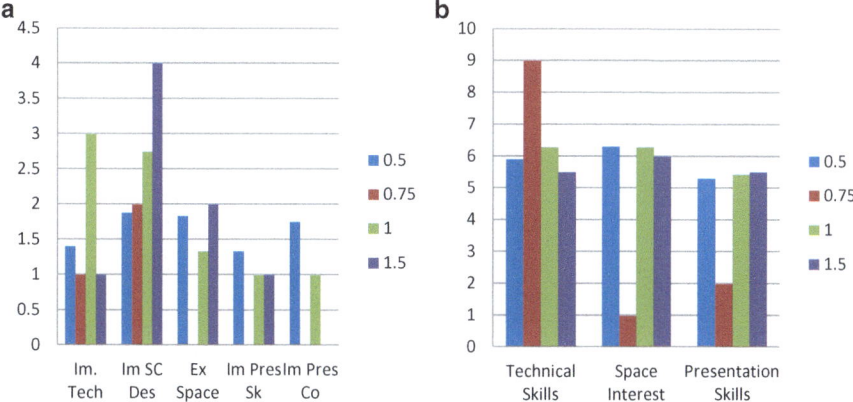

Fig. 10.22 (**a**) Improvement in status levels, compared by time participating (in academic years). (**b**) Effect of program on causing improvement by status, compared by time participating (in academic years)

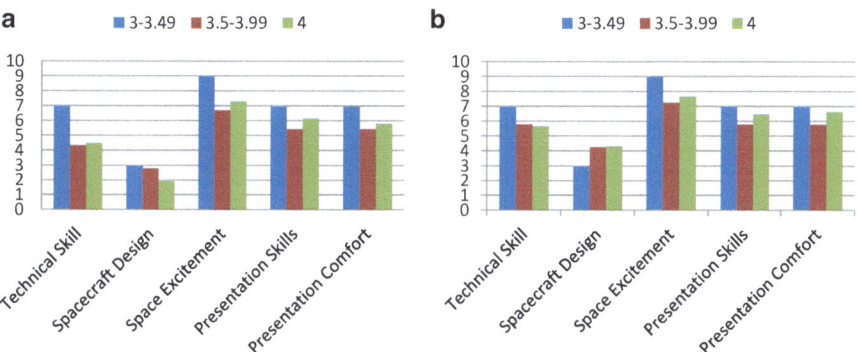

Fig. 10.23 (**a**) Beginning status levels, compared by participant GPA. (**b**) Ending status levels, compared by participant GPA

10.3.9 Comparison of Results by GPA

This section compares the various success indicators and the GPA of the participants in an attempt to determine whether there is any correlation. Figure 10.23a, b presents the pre-participation and post-participation status levels. There does not appear to be, as expected, any strong bias towards or away from certain categories which correlates with GPA. Figure 10.24a shows the average aggregate improvement. This indicates a slight improvement which correlates with increased GPA (5.67 vs. 6.2). Again, excluding the individual who indicated limited participation causes the 3.49–4.00 to overtake the 4.0 GPA condition (increasing it to 6.38). The other data point (where improvement is attributed, but none is shown) is the sole member of the 3.0–3.49 condition, so this has no impact on the 4.0 versus 3.49–4.00

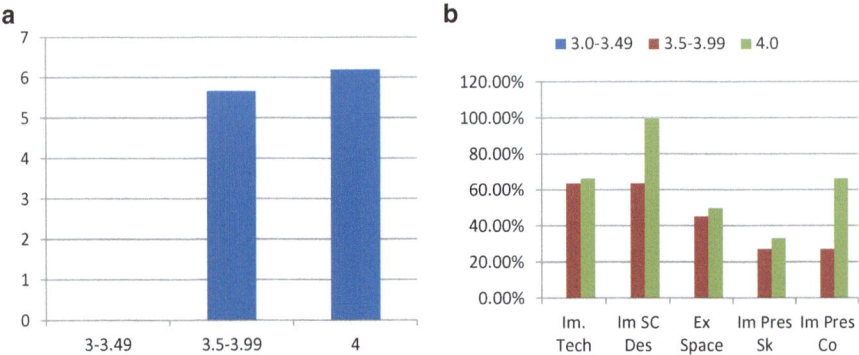

Fig. 10.24 (**a**) Average aggregate improvement, compared by participant GPA. (**b**) Percentage of participants showing improvement in each status, compared by participant GPA

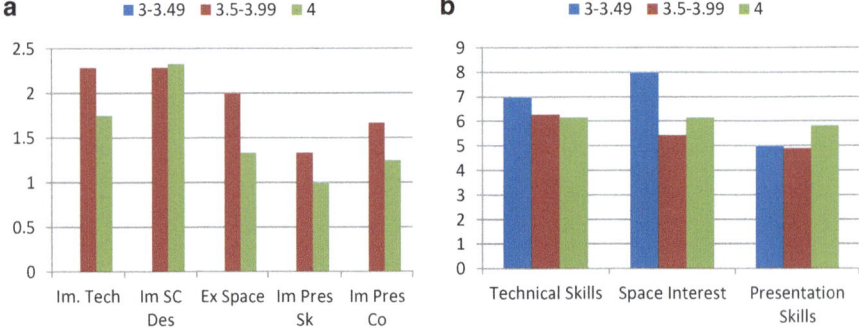

Fig. 10.25 (**a**) Improvement in status levels, compared by participant GPA. (**b**) Effect of program on causing improvement by status, compared by participant GPA

comparison. Figure 10.24b indicates a higher percentage of individuals in the 4.0 condition experienced an increase in each category as compared to the 3.5–3.99 condition. Excluding the aforementioned individual causes the 3.5–3.99 to overtake in one area (technical skills) and match in another (excitement about space).

The average amount of improvement values, shown in Fig. 10.25a, shows that the 3.5–3.99 category experienced more improvement when improvement occurred, in all but one category (spacecraft design). Results in the attribution responses shown in Fig. 10.24b are mixed with the 3.5–3.99 condition scoring higher in one (technical skills) and the 4.0 condition scoring higher in the other two. The 3–3.49 condition outscores the other two in two conditions (outscoring 3.5–3.99 in all three); however, as there is only a single member to this condition there is insufficient evidence of anything significant about this. Excluding the previously discussed data point does not impact these results.

From the aforementioned, there is insufficient evidence to conclude that GPA had any particular correlation with the level of value gained from program participa-

10.3 Formative Evaluation Tools: Gain Quantification Survey

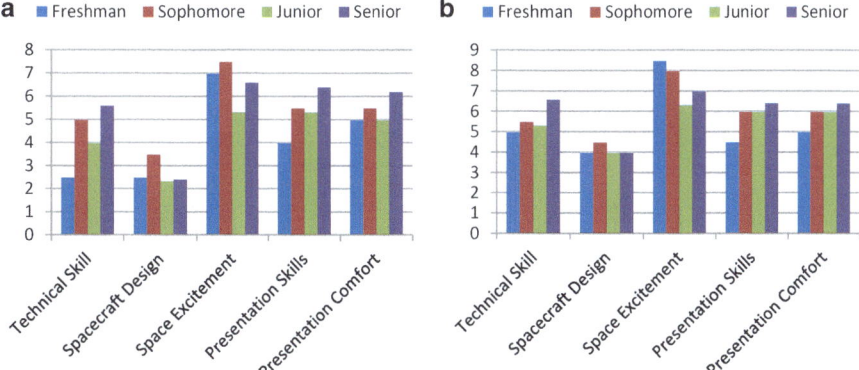

Fig. 10.26 (**a**) Beginning status levels, compared by undergraduate class level. (**b**) Ending status levels, compared by undergraduate class level

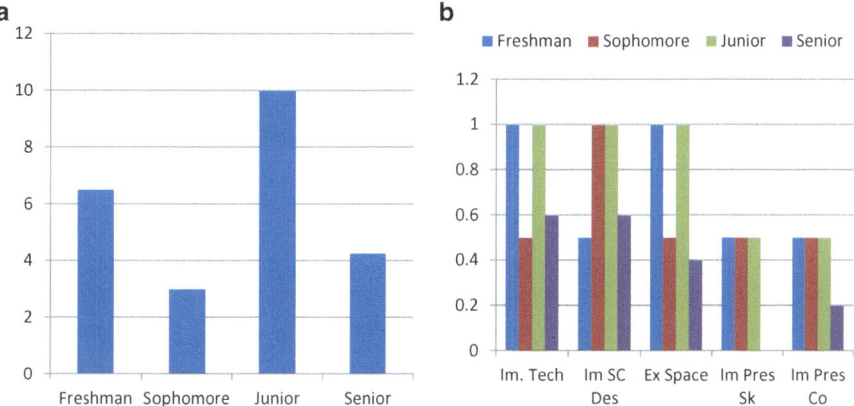

Fig. 10.27 (**a**) Average aggregate improvement, compared by undergraduate class level. (**b**) Percentage of participants showing improvement in each status, compared by class level

tion, as indicators conflicted. Moreover, in the areas where one was shown to outperform another, there is no practical significance to the result.

10.3.10 Comparison of Results by Undergraduate Class Level

The final area of consideration is to determine whether a correlation exists between undergraduate class level (freshman, sophomore, junior or senior) and results. Figure 10.26a, b shows the initial and ending status levels. Figure 10.27a, b shows a lack of progressive increase with increase in grade in aggregate improvement and percentage of individuals improving in each category. Finally, Fig. 10.28a, b shows

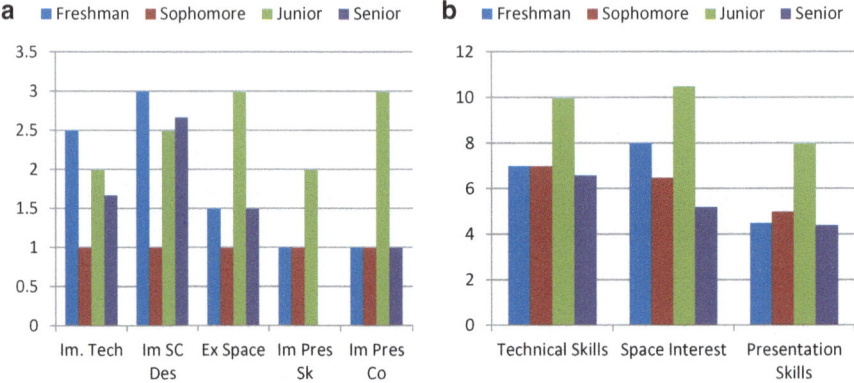

Fig. 10.28 (**a**) Improvement in status levels, compared by undergraduate class level. (**b**) Effect of program on causing improvement by status, compared by undergraduate class level

a lack of progressive correlation in the level of improvement experienced and attribution of improvement to the program.

10.3.11 Summary

This section has presented assessment work related to the University of North Dakota's OOSDI. It has demonstrated benefit from participation in all of the categories of learning objectives identified prior to program initiation (and several not explicitly identified). It has also shown a strong correlation between the level of improvement and participation as a team lead. It has also shown strong correlation between the number of hours per week that individuals participated and their average aggregate improvement. Similarly, a correlation between the duration of participation and improvement is shown. No significant confounding correlation is shown between graduate versus undergraduate status, participant GPA, and undergraduate class level and the level of improvement shown (with conflicting indicators or a lack of progression shown).

10.4 Formative Evaluation Results and Program Enhancements in Prior Work

The foregoing formative evaluation techniques have been used to adjust the focus of OOSDI throughout its existence, allowing each semester's activities to jointly focus on the advancement of the overall mission concept and spacecraft, as well as on the identified goals and needs of program participants. The data from the gain quantification survey has been used to adjust program participation styles and activities towards those approaches that increase student educational outcomes on a semester-by-semester basis.

10.5 Discussion of the Use of Summative Evaluation

In contrast to formative assessment (or evaluation), summative evaluation seeks to assess the work that has already been conducted to facilitate its scholarly dissemination, program performance assessment, student outcome assessment, and for other purposes. While formative assessment seeks to improve outcomes, summative evaluation, thus, seeks to characterize which and how well particular outcomes were achieved. Overlap none-the-less exists, as summative evaluation for one program may inform the development and nature of other programs or, as was the case with OOSDI, summative evaluation data from one portion of the program was used to have a formative impact on subsequent portions.

Subsequent sections provide a discussion of the dual-mode gain quantification survey instrument, from a summative perspective. Then, the Undergraduate Research Student Self-Assessment instrument (also used for summative evaluation) is presented.

10.6 Summative Evaluation Tools: Gain Quantification Survey

As was discussed in Sect. 10.3, five key areas of focus were initially identified when attempting to ascertain the educational effectiveness of the OOSDI. These were technical skills in the area of focus, spacecraft design skills, space excitement, and skills and comfort of giving presentations. Additional areas of focus have subsequently been identified; however, due to limited data, they are not considered at this time.

Initially, a 26-question survey [3] (described in Sect. 10.3) was conducted. The first 12 questions collected demographic data about the participants and their involvement in OOSDI. This included information about their academic status (undergraduate vs. graduate, class, time in program, GPA). They were also asked about their participation in OOSDI (number of years participating, hours participating per week, and whether they participated for academic credit or not) and whether they had previous involvement in spacecraft design and, if so, the type of involvement.

The second version of this survey asked 42 questions, including the 26 from the initial survey. The duration of participation was changed to semesters from years to avoid causing students' confusion in answering this question. The data from the earlier survey was multiplied by two (as UND has two normal year semesters and the program, at this time, had extremely minimal participation by students over the summer months) to allow the year and semester numbers to be utilized together. This treatment is consistent with the clarification provided regarding how to answer the previous year's question. An additional option was added to the question regarding academic participation and the questions regarding whether the participant had received credit from participating and the type of for-credit participation were combined into a single question (including answer in the format "yes-type" and the answer "no").

The majority of the survey comprises questions presented on a nine-point scale. In all cases 9 is the superior answer (indicating greater experience/knowledge/etc.) 5 is neutral, and 1 indicates the inferior answer. Students were asked, on both surveys, to characterize their pre-participation status and current status with regards to each of the metric areas.

The summative value of this survey is demonstrated by the use of data from these two survey forms to assess the correlation between the duration of participation (in semesters) and the level of benefit attained. This level of benefit has been calculated, in all cases, by subtracting the pre-participation status level from the post-participation status level. In a very limited number of cases, this resulted in inexplicable negative values. These have been investigated (as discussed in [3]) and replaced with zeroes, as they appear (based on responses to the attribution questions) to represent clerical errors by participants; they are, otherwise, uninterpretable.

10.6.1 Results and Discussion

This section presents the process and results of the analysis of the data that was collected as described in the foregoing section. In addition to the five key metrics for which data was collected, an additional aggregate metric was created by adding the benefit attained from each of the other areas together. As a participant would not necessarily receive benefit in all areas, this combined metric may serve as more holistic view of the value of the program to the participant. This analysis process commenced with the correlation between all respondents' performance in each of the five key metrics and the combined aggregate metric. This data, which is based on 31 respondent surveys including nine master's students, twenty-one undergraduate students, and one individual who did not respond as to his or her student's status, is presented in Table 10.1. This data included responses from seventeen non-lead participants, thirteen lead level participants, and one individual who did not respond with regards to this question.

From the data presented, it is clear that there is limited correlation between the metrics and the amount of time spent participating, when neglecting all other confounding factors. For four of the metrics, correlation levels of between 0.17 and 0.26 are reported; one (spacecraft design) reports a negative correlation level. The aggregate improvement correlation level is also below 0.25. As correlation levels are on a -1 to 1 scale (with -1 indicating perfect inverse correlation and 1 indicating perfect correlation), these values do not provide much support to the thesis that prolonged participation provides greater levels of benefit.

Because of the nature of student participation, however, some students will receive benefits at different rates. Possible confounding variables are now considered. In Table 10.2, the correlation process is separated into graduate or undergraduate status.

For the master's students, this data shows strong correlation in three areas (technical skills, spacecraft design, and aggregate improvement); moderate correlation is also shown in presentation skills. The level of excitement and presentation comfort

10.6 Summative Evaluation Tools: Gain Quantification Survey

Table 10.2 Comparison of the correlation between number of semesters involved in the program and the five assessed metrics, for master's and undergraduate students

	Technical skill	Spacecraft design	Level of excitement	Presentation skills	Presentation comfort	Aggregate improvement
Bachelor's students	−0.12	−0.21	−0.24	0.10	0.04	−0.17
Master's students	0.85	0.84	0.35	0.63	0.37	0.89

Table 10.3 Comparison of the correlation between the number of semesters involved in the program and the five assessed metrics between those serving in a participant and team lead role

	Technical skill	Spacecraft design	Level of excitement	Presentation skills	Presentation comfort	Aggregate improvement
Participant	−0.09	−0.19	−0.23	0.28	0.21	−0.11
Team lead	0.72	0.63	−0.03	0.13	0.21	0.52

Table 10.4 Comparison of the correlation between the number of semesters involved in the program and the five assessed metrics between those devoting between 1 and 3.99 hours per week and those devoting 4 and 7.99 hours per week

	Technical skill	Spacecraft design	Level of excitement	Presentation skills	Presentation comfort	Aggregate improvement
1–3.99	0.30	0.08	−0.27	0.36	0.39	0.27
4–7.99	0.33	0.32	−0.05	0.04	−0.24	0.18

metrics show greater correlation levels than with the non-separated data. However, for the bachelor's students, limited positive and negative correlation data is still presented. For the purposes of assessing statistical significance, an improvement as a function of duration of participation value was created by dividing each value by the duration (in semesters) of participation. From this, the difference in terms of space excitement (0.04) was significant at $p=0.05$ and technical skills (0.08) and aggregate improvement (0.07) were significant at $p=0.10$. Spacecraft design (0.18), presentation skills (0.12), and presentation confidence (0.48) were not shown to be significantly different at either $p=0.05$ or $p=0.10$.

Next, the correlation assessment is performed separating participants and team leads. This data is presented in Table 10.3.

Team leads, overall, show moderate correlation in technical skills, spacecraft design, and aggregate improvement and limited correlation in presentation skills and comfort. More pronounced results are demonstrated by the master's level students. Very limited negative correlation is shown for the level of excitement. Undergraduate participants, however, show limited positive or negative correlation for all metrics, while the very limited data set for master's level participants shows perfect correlation.

As the amount of weekly involvement in the project may affect the level of benefits obtained, the five metrics are now correlated separated by the number of hours worked each week (one 8+ response has been removed as it represented insufficient data for analysis). This data is presented in Table 10.4.

While this segmentation produces five combined correlation values in the 0.3–0.4 range and three in the ±0.2–0.3 range, no clear trend emerges for the combined data. The master's students, however, show strong correlation in the majority of the metrics. The limited set of undergraduate high commitment data appears to show a pronounced negative correlation; however, this is likely attributable to students entering and leaving this group between survey administrations (instead of the unrealistic conclusion that students were unlearning skills). It also may be indicative of students reevaluating their own skill levels in light of a better understanding of the subject material.

Next, the data was correlated segmented by whether the participants received academic credit for their participation or not. This data is presented in Table 10.5. Again, several combined correlation values in the 0.3–0.4 range are produced and one value in the 0.4–0.5 range is generated. Two values in the ±0.2–0.3 range are also indicated. However, no moderate or strong correlation values are indicated. Insufficient master's level for-credit participants were present to allow reporting; however, the impact of this limited set on the combined metric is pronounced. There is also a clear difference between the master's and bachelor's level students in the no-credit category.

The majors of the participants may also have some impact on the correlation between the duration of participation and the level of benefit received. As the project is run out of the UND Computer Science Department, there has been consistent computer science student involvement throughout the project. Thus, correlation is now performed for only computer science students. This data is presented in Table 10.6.

Again, no strong trends are present. One correlation value in the 0.3–0.4 range and two in the ±0.2–0.3 range are produced. Finally, the correlation segmented by both the field of major (divided into computer science and others) and academic level (graduate or undergraduate) is calculated. This data is presented in Table 10.7.

Table 10.5 Comparison of the correlation between the number of semesters involved in the program and the five assessed metrics between those participating for course credit and those not participating for course credit

	Technical skill	Spacecraft design	Level of excitement	Presentation skills	Presentation comfort	Aggregate improvement
Credit	0.38	0.30	−0.04	0.19	−0.25	0.23
No credit	0.03	−0.14	−0.12	0.12	0.43	0.10

Table 10.6 Correlation between number of semesters involved in the program and the five assessed metrics, for computer science students

	Technical skill	Spacecraft design	Level of excitement	Presentation skills	Presentation comfort	Aggregate improvement
Computer Science	0.31	0.14	−0.21	0.18	0.17	0.20

Table 10.7 Correlation between the number of semesters participating and increase in the five assessed metrics, for computer science and non-computer science students at both the bachelor's and master's levels

	Technical skill	Spacecraft design	Level of excitement	Presentation skills	Presentation comfort	Aggregate improvement
Bachelor's: computer science	−0.07	−0.22	−0.27	0.08	0.01	−0.17
Master's: computer science	0.95	0.88	0.07	0.42	0.34	0.95
Bachelor's: other	−0.95	−0.90	N/A	N/A	N/A	−1.00
Master's: other	0.90	0.89	0.58	N/A	0.58	0.86

This data indicates very strong correlation for both computer science and non-computer science master's level students in the technical skill, spacecraft design and aggregate improvement categories. Moderate correlation is shown for non-computer science students in level of excitement and presentation comfort. The master's level computer science students also produce one correlation value in the 0.3–0.4 range and one in the 0.4–0.5 range. Note that in several cases, data characteristics generated a divide-by-zero issue for the excel correlation function. These are indicated with a "N/A" in Table 10.7.

What is problematic in this data is the very strong negative correlation values reported for non-computer science undergraduates (−0.95, −0.9, and −1). As it seemed unlikely that this could be attributable to the nature of the program (e.g., students are gaining less value the longer they spend), this was investigated. This phenomenon was caused by there being only a limited number of respondents. Some of these respondents achieved a significant level of gain in a single semester, while others achieved moderate levels of improvement over a longer period of time. This result is, thus, a quirk of the data and limited number of respondents in this category.

10.6.2 Summary

This section has demonstrated the use of the gain quantification survey for summative analysis by considering the correlation between the duration of participation in the program and the level of value attained. For several groups of participants, this correlation was shown to exist. However, it was not shown to exist in the general case (attributable, as demonstrated, to the presence of confounding variables). There were also some groups where no correlation between benefit level and participation duration could be demonstrated. For graduate students, a very strong correlation level was shown to exist in the technical and spacecraft design skills categories. Strong correlation was shown when considering the master's students as a group and this was even stronger when they were divided into computer science and non-computer science students. Presentation skills was shown to have a moderate level of correlation as a combined (all master's students) group. This correlation couldn't be assessed separately for the non-computer science master's students and was not as pronounced when only the computer science master's students were considered. Level of excitement and presentation comfort showed moderate correlation for the non-computer science master's students and a lower level of correlation for the computer science master's students. Both the computer science and non-computer science master's students showed a strong (0.95 in the case of computer science master's students) aggregate improvement correlation; this was also evidenced in the combined (all master's students) data.

Team leads (which included both graduate and undergraduates) showed moderate correlation between the duration of participation and level of value attained in three areas (technical and spacecraft design skills) and aggregate improvement. However, beyond the team leads, no grouping could be found from the data elements collected that removed a confounding variable to demonstrate strong positive correlation between the duration of participation and the level of benefit attained. This could indicate that undergraduates do not gain significantly in additional benefit beyond some level of participation. It could also be a function of limited sample sizes. Perhaps there is some other confounding variable that was not identified or surveyed that, if the data was segmented by it, would allow this correlation to be demonstrated.

The presence or absence of correlation is important for several reasons. First, the presence of correlation suggests the value of longer duration participation. This correlation was an expected outcome. Second, when data shows a significant difference when segmented over a variable, this identifies the variable as a prospective factor in the level of benefit that students attain. Some of these (such as team lead status, for-credit participation status, or the number of hours worked a week) can be controlled if greater levels of benefit are seen for one status than others. Others (such as the major or graduate vs. undergraduate status) demonstrate what groups should be targeted for participation, as students in these groups receive particular benefit. The identification of groups (of all types) that do not perform as well as others also serves to focus attention on improving (or identifying non-modifiable limiting factors) the outcomes of the program for these groups.

10.7 Summative Evaluation Tools: URSSA

This section describes the Undergraduate Research Student Self-Assessment (URSSA) instrument as well as its use in assessing a small spacecraft development program. The URSSA is a disciple-agnostic instrument that was designed to assess the value provided from research program participation, on a summative basis, to undergraduates. The instrument characterizes students' attitudes and beliefs regarding their experience as well as assessing whether a number of common products of research (e.g., papers) have been generated by the student. The URSSA also characterizes the impact of research participation on students' desire to attend graduate (or professional) school.

10.7.1 Undergraduate Research Student Self-Assessment Mechanism

The URSSA [10] mechanism is a highly validated [11], widely used assessment for quantifying the benefits of research participation by undergraduate students. The University of Colorado at Boulder team that developed the URSSA conducted an 8-year-long study of undergraduate research at multiple institutions comprising over 350 interviews. They conducted three evaluations of undergraduate research programs (including an additional 350 interviews and surveying 150 students). Finally, they performed a literature review with regards to relevant studies of undergraduate research evaluation.

From this, they developed an initial survey, refined this with the so-called think-aloud interviews to assess interpretation of the questions' wording and conducted a pilot study including more than 500 students at 24 institutions. They used confirmatory factor analysis and removed or changed items as necessary to correct remaining issues.

10.7.2 Experiment Implementation

The URSSA survey was given (without previous announcement which could result in self-selection to take/not take the survey) at final group meetings to all of the participants who were at each meeting. Two graduate students who commenced participation as an undergraduate and continued their participation as a graduate student were included in those surveyed.

Data related to 14 interrelated areas of focus was collected by the URSSA survey. The majority of this data (excluding sections that were not relevant to this inquiry) is now presented.

The first area of focus related to areas of gain. Students were asked to indicate "how much did you GAIN in the following areas as a result of your participation in this program" in response to eight areas:

1.1 Analyzing data for patterns.
1.2 Figuring out the next step in a research project.
1.3 Problem solving in general.
1.4 Formulating a research question that could be answered with data.

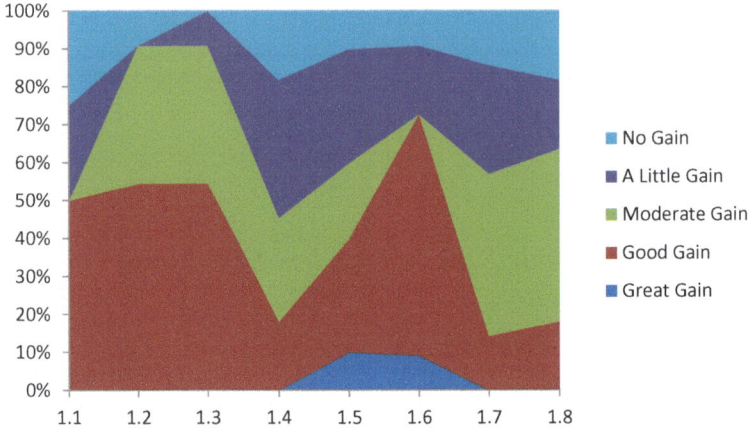

Fig. 10.29 Skills gained

1.5 Identifying limitations of research methods and designs.
1.6 Understanding the theory and concepts guiding my research project.
1.7 Understanding the connections among scientific disciplines.
1.8 Understanding the relevance of research to my coursework.

They were able to select from the choices of no gain, a little gain, moderate gain, good gain, great gain, and not applicable. Figure 10.29 presents the number of individuals responding in each category (those indicating 'not applicable' were excluded from the percentage calculation). Additionally, for all results presented in this section, the results of an individual who advised of a personal issue that biased responses and an individual who indicated virtually no participation were excluded, so as to allow focus on those who participated.

Notably, over 70 % of respondents indicated some level of gain in each category. While the 'great gain' responses were minimal, 'good' and 'moderate' gain both had large groups indicating them for most categories. Most of the respondents enjoyed gain in most of these categories and this was largely gain in the 'good' and 'moderate' categories.

The next area of focus dealt with confidence and comfort-related topics. These responses are indicated in Fig. 10.30. Students were again given the choice of no gain, a little gain, moderate gain, good gain, great gain, and not applicable. They also, again, had eight areas to respond to:

2.1 Confidence in my ability to contribute to science.
2.2 Comfort in discussing scientific concepts with others.
2.3 Comfort in working collaboratively with others.
2.4 Confidence in my ability to do well in future science courses.
2.5 Ability to work independently.
2.6 Developing patience with the slow pace of research.
2.7 Understanding what everyday research work is like.
2.8 Taking greater care in conducting procedures in the lab or field.

10.7 Summative Evaluation Tools: URSSA

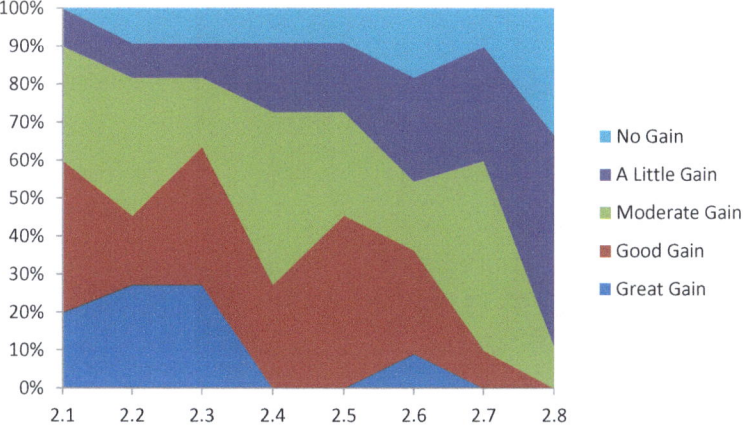

Fig. 10.30 Confidence, comfort, and skills gained

Again, the data can be characterized as showing that most respondents enjoyed benefit in most areas. All respondents indicated at least a little gain in "2.1 Confidence in my ability to contribute to science," with 90 % indicating at least moderate gain in this category. For "2.2 Comfort in discussing scientific concepts with others," "2.3 Comfort in working collaboratively with others," "2.4 Confidence in my ability to do well in future science courses," "2.5 Ability to work independently," and "2.7 Understanding what everyday research work is like," 90 % reported at least a little gain in this area and more than 70 % indicated at least moderate gain. Between 20 % and 30 % of respondents reported great gain in the areas of "2.1 Confidence in my ability to contribute to science," "2.2 Comfort in discussing scientific concepts with others," and "2.3 Comfort in working collaboratively with others" from their participation.

Focus area 3 dealt with the performance and dissemination of research. These results are presented in Fig. 10.31. Students were, again, given the choice of no gain, a little gain, moderate gain, good gain, great gain, and not applicable. In this case, they had 13 areas to respond to:

3.1 Writing scientific reports or papers.
3.2 Making oral presentations.
3.3 Defending an argument when asked questions.
3.4 Explaining my project to people outside my field.
3.5 Preparing a scientific poster.
3.6 Keeping a detailed lab notebook.
3.7 Conducting observations in the lab or field.
3.8 Using statistics to analyze data.
3.9 Calibrating instruments needed for measurement.
3.10 Working with computers.
3.11 Understanding journal articles.
3.12 Conducting database or Internet searches.
3.13 Managing my time.

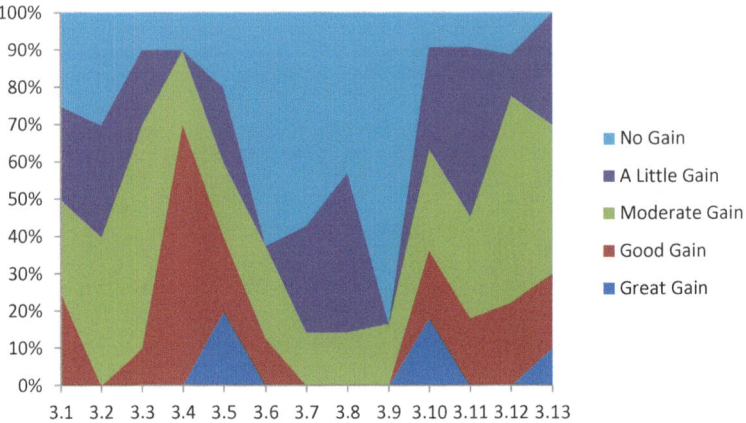

Fig. 10.31 Skills and abilities gained

The results in response to this question varied more than to the previous areas of focus. Less than 50 % of respondents reported gain in the categories of "3.6 Keeping a detailed lab notebook," "3.7 Conducting observations in the lab or field," and "3.9 Calibrating instruments needed for measurement." This is not unexpected as there was little stress placed on lab notebook use in most groups (a version control management system, more typical of industry than academia, was used to manage software and design documents; this required participants to enter what they had done for each update, establishing individual contributions). Focus was also not placed on observations or the use of instrumentation, though a limited group of students participated in the thermal vacuum testing of components and had some exposure (as indicated by the limited number of respondents indicating gain in these areas).

At least 70 % of respondents reported some gain in the areas of "3.1 Writing scientific reports or papers" and "3.2 Making oral presentations" and approximately 90 % indicate gain in "3.3 Defending an argument when asked questions" and "3.4 Explaining my project to people outside my field" (in fact 90 % indicated at least moderate gain in response to 3.4 and nearly 70 % indicated good gain to this question, a clear indication of the benefit of the interdisciplinary nature of the program in this area). Over 80 % of respondents indicated some gain in "3.5 Preparing a scientific poster" and over 60 % indicated at least moderate gain in this category. At least 90 % of respondents indicated gain in "3.10 Working with computers," "3.11 Understanding journal articles," and "3.12 Conducting database or Internet searches" and 100 % indicated some gain in "3.13 Managing my time," all skills that will be valuable in both continued academic pursuits as well as in a research or industry workplace.

Focus area 4 asked students to characterize their attitudes towards the project. The responses to these questions are presented in Fig. 10.32. In this instance, student response choices included none, a little, some, a fair amount, a great deal, and not applicable. They had eight areas to respond to:

10.7 Summative Evaluation Tools: URSSA

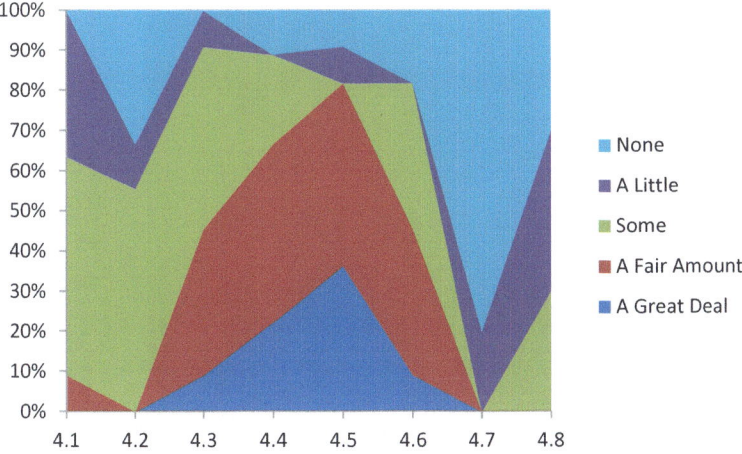

Fig. 10.32 Student Attitudes

4.1 Engage in real-world science research.
4.2 Feel like a scientist.
4.3 Think creatively about the project.
4.4 Try out new ideas or procedures on your own.
4.5 Feel responsible for the project.
4.6 Work extra hours because you were excited about the research.
4.7 Interact with scientists from outside your school.
4.8 Feel a part of a scientific community.

This data can again be characterized with the statement that most respondents had some level of this positive attitude in most cases. The main exception to this is "4.7 Interact with scientists from outside your school." This is again not unexpected as there was limited opportunities for this type of interaction during this period (though more in this category would be expected during integration and launch preparation activities, where extensive coordination with the spacecraft-into-rocket integrators and launch provider will be expected). The attendance of more students at conferences, currently limited by budgetary factors and student availability due to other commitments, would also potentially stand to increase this. All respondents indicated at least 'a little' feeling in response to the statements "4.1 Engage in real-world science research" and "4.3 Think creatively about the project" (with 90 % indicating some or higher level of this feeling in response to 4.3), clearly demonstrating the benefit of the project's real-world nature and supporting Ayob et al.'s [12] assertion of the creativity benefits of PBL.

The fifth area of focus relates to the research experience and mentor relationship. Students were given the choices of not applicable, poor, fair, good, and excellent in response to these statements. They were asked to respond (the responses are depicted in Fig. 10.33) in regards to:

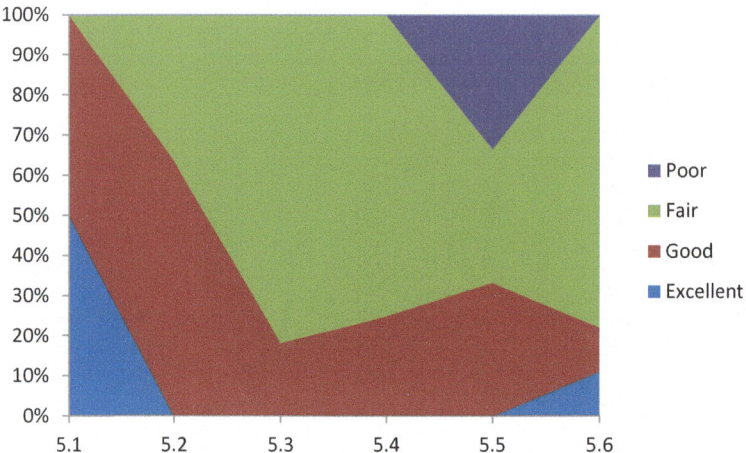

Fig. 10.33 Respondent attitudes

5.1 My working relationship with my research mentor.
5.2 My working relationship with research group members.
5.3 The amount of time I spent doing meaningful research.
5.4 The amount of time I spent with my research mentor.
5.5 The advice my research mentor provided about careers or graduate school.
5.6 The research experience overall.

It is important to note that the term "mentor" is used frequently in the project to refer to the faculty mentors for each group, even though a lot of mentorship is actually performed by more senior students. For this reason, this data may not fully capture the mentorship attitudes of respondents.

All respondents rated their attitude as at least fair in response to all categories except "5.5 the advice my research mentor provided about careers or graduate school." A focus was never placed on this, so this is not unexpected. The availability of this information or time devoted to it has not been identified as a focus of attention in the past; however, assessment of whether this is an expectation or desire of participants will serve as a subject for future work. All of the respondents indicated that their relationship with their research mentor was at least good and 50 % indicated it was excellent. Approximately 65 % of respondents indicated that their relationship with group members was good, with the rest indicating it was fair.

Further research is ongoing with regards to "5.3 the amount of time I spent doing meaningful research" and "5.4 the amount of time I spent with my research mentor." The research mentor interaction time is an area of previous concern and faculty mentors have been encouraged to be more active where possible. A time study will be conducted to see why participants are not rating the amount of time "doing meaningful research" as at least good, to determine whether they desire greater levels of participation or feel that time is being spent on nonmeaningful activities. It may also be important to set expectations with regards to any unresolvable

10.7 Summative Evaluation Tools: URSSA

Table 10.8 Student accomplishments

		Yes (%)	No (%)
6.1	I presented a talk or poster to other students or faculty	45	55
6.2	I presented a talk or poster at a professional conference	18	82
6.3	I attended a conference	18	82
6.4	I wrote or cowrote a paper that was published in an academic journal	18	82
6.5	I wrote or cowrote a paper that was published in an undergraduate research journal	9	91
6.6	I will present a talk or poster to other students and faculty	64	36
6.7	I will present a talk or poster at a professional conference	10	90
6.8	I will write or cowrite a paper to be published in an academic journal	10	90
6.9	I will write or cowrite a paper to be published in an undergraduate research journal	10	90
6.10	I won an award or scholarship based on my research	0	100

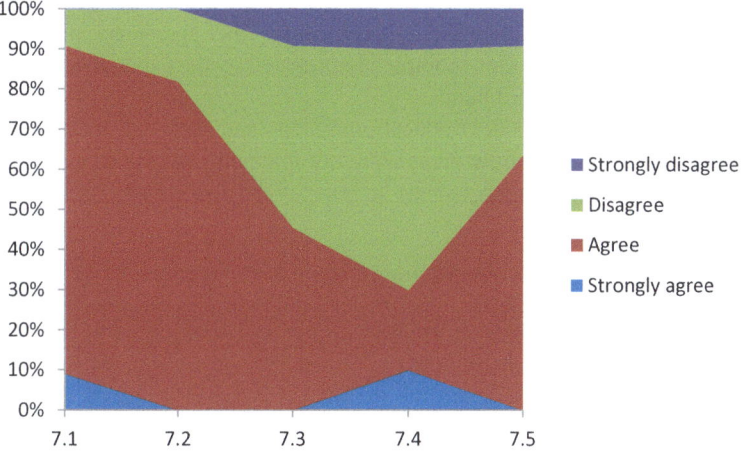

Fig. 10.34 Preparation benefits of participation

discrepancies in this regard and to help students see why their activities (which they may not see the contribution of) are in fact meaningful. This may also improve "5.6 the research experience overall."

The next focus area deals with the specific accomplishments of students in the program. Table 10.8 presents the percentage of students indicating accomplishment in each area. It is important to note that many participants may experience these benefits later in the course of their participation in the project.

The next URSSA section assesses student preparation benefits for their next educational steps. This data is presented in Fig. 10.34. Unlike many users of the URSSA that may seek to attract primarily students interested in pursuing graduate education and research careers, many students in the OpenOrbiter program are planning to go into industry and hope to (and generally do) gain real-world, hands-on experience from participation that is relevant to this goal from project participation. Students were asked to respond to:

7.1 Doing research confirmed my interest in my field of study.
7.2 Doing research clarified for me which field of study I want to pursue.
7.3 My research experience has prepared me for advanced coursework or thesis work.
7.4 My research experience has prepared me for graduate school.
7.5 My research experience has prepared me for a job.

In the questions that were both industry and academic/research-career relevant, high responses were recorded including 90 % indicating agreement or strong agreement to "7.1 Doing research confirmed my interest in my field of study" and 80 % indicating agreement to "7.2 Doing research clarified for me which field of study I want to pursue." The results for 7.3 and 7.4 were noticeably lower. Approximately 65 % agreed that the research experience has prepared them for a job. This question is, however, somewhat ambiguous as it is unclear whether agreement indicated that the respondent felt that they were completely prepared by the experience or just that the experience helped prepare them.

Section 10.8 focused on next educational steps and the program's impact on decision making. Many of these areas were not relevant (e.g., medical/dental/professional degree) to the undergraduate majors of the individuals. Students were asked to respond to:

8.1 Enroll in a Ph.D. program in science, mathematics, or engineering?
8.2 Enroll in a master's program in science, mathematics, or engineering?
8.3 Enroll in a combined M.D./Ph.D. program?
8.4 Enroll in medical or dental school?
8.5 Enroll in a program to earn a different professional degree (i.e., law, veterinary medicine)?
8.6 Pursue certification as a teacher?
8.7 Work in a science lab?

Approximately 50 % indicated that they were at least a little more likely (and 40 % indicate somewhat more likely) to pursue a master's degree due to participation. Approximately 35 % indicated that they were at least a little more likely to pursue a Ph.D. and work in a science lab, due to participation.

The academic level of undergraduate respondents is presented in Table 10.9. This distribution corresponds roughly with key semesters of aggressive versus less

Table 10.9 Academic year of respondents

Year	Students (%)
Senior	22
Junior	44
Sophomore	33

10.7 Summative Evaluation Tools: URSSA

aggressive recruiting as well as showing a gradual buildup of participants over time (leading to the higher percentages of juniors and seniors).

For the purpose of completeness of reporting, Table 10.10 presents the responses to the questions about how respondents found out about the program. It is notable that those who were recruited (and thus were respondents) were likely made aware of it in multiple ways. All respondents learned about the project in class; all respondents indicated at least two ways of awareness generation (and in many cases more).

Table 10.11 presents the reasons for participant participation. Notably a desire to learn about graduate school and pursuit of a science research career were not motivators for most individuals' involvement. No respondents indicated that they participated to work with a particular faculty member and the program's reputation and recommendation letters were not seen as key drivers either. The number of hours spent by participants, on average, is reported in Table 10.12.

Table 10.10 Sources of information about the program

		Yes (%)	No (%)
12.1	I knew this institution offered research opportunities before coming here	18	82
12.2	In class	100	0
12.3	An academic advisor	45	55
12.4	An announcement (flyer, poster, e-mail, Web site, etc.)	82	18
12.5	A presentation given by professors or students about their research	82	18

Table 10.11 Reasons for participation

		Yes (%)	No (%)
13.1	Explore my interest in science	91	9
13.2	Gain hands-on experience in research	91	9
13.3	Clarify which field I wanted to study	91	9
13.4	Clarify whether graduate school would be a good choice for me	27	73
13.5	Clarify whether I wanted to pursue a science research career	27	73
13.6	Have a good intellectual challenge	91	9
13.7	Work more closely with a particular faculty member	0	100
13.8	Participate in a program with strong reputation	27	73
13.9	Get good letters of recommendation	36	64
13.10	Enhance my resume	91	9

Table 10.12 Number of hours spent

Number of hours	Percent (%)
1–5	64
6–10	36

From the foregoing, it is clear that many different types of benefit were obtained by participants. Prior to conducting this survey, some of these areas of benefit were not identified as focus areas; however, benefit accrued in some of these areas despite this.

10.7.3 Summary

This section has presented a pedagogical overview of and data collected using the URSSA assessment tool about the OOSDI. It has highlighted a pathway for others interested in forming a small spacecraft development program at their institution to do so, using the presented evidence of the educational benefits attained from using this type of program for STEM student learning. The foregoing indicates that most respondents gained benefits in many areas from project participation. It also indicated a small number of areas for future focus of improvement. The free-response questions also provide anecdotal support for this, with respondents stating that they are "likely to enroll" in a graduate program because of participation, that participation increased their confidence with regards to the ability to do research, and that "it has helped me gain an understanding of a possible field I could focus on" and "it exposed me to new areas of computer science that I am considering focusing on for a career."

10.8 Summative Evaluation Tools: Meeting Expectations Survey

The experimental design for the pre-participation survey was discussed previously in this chapter and in [1]. A post-participation survey, which followed this same design (except that students were asked to identify the benefits received instead of expected), was also conducted. The pre-participation survey was, logically, administered at the beginning of a semester and the post-participation survey was administered at the end. As is typical with extracurricular enrichment activities, significant attrition occurred during the semester. For both the pre- and postsurvey, the instrument was administered to all individuals attending the relevant meetings during the week of administration (and these attendance levels were not abnormal). With the post-participation survey, two respondents' responses have been excluded. The first was excluded due to the indicated lack of participation during the semester (making the data of little value in assessing the benefits of participation and a candidate for removal as an outlier, even without the provided explanation by the respondent). The second exclusion was attributable to a disclosed respondent personal issue.

Survey respondents included both graduate and undergraduate students. In the pre-participation survey, eighteen undergraduates and four graduate students responded. In the post-participation survey, responses were received (after exclusions) from nine undergraduate and two graduate students. The class membership of

10.8 Summative Evaluation Tools: Meeting Expectations Survey

Table 10.13 Class status of undergraduate respondents

	Presurvey	Postsurvey
Senior	6	2
Junior	7	4
Sophomore	5	3
Freshman	1	0

Table 10.14 Student respondents for pre- and post-participation surveys, by major and minor

	Air traffic control	Electrical engineering	Theat.	Info systems	Math	Comp. Sci.	Phil.	Poli. Sci.	Crim. justice
Major: initial	1	1	0	0	0	21	0	0	0
Minor: initial	0	0	1	1	8	1	1	1	0
Major: post	0	1	0	0	0	10	0	0	0
Minor: post	0	0	0	0	6	1	0	0	1

the respondents to both the pre-participation and post-participation surveys is indicated in Table 10.13.

The respondents to the survey were predominantly computer science students, due to the then-current software development focus of the project. Table 10.14 presents the majors and minors of the respondents for both the pre- and post-participation surveys.

Both surveys were conducted anonymously and collected the demographic information (discussed above) in addition to data (presented in Sect. 10.4) regarding their expectations or benefits received. In addition to being asked about their major, minor, graduate versus undergraduate status and class level, students were also asked whether they had previously participated in the OOSDI or not. In the initial survey, 12 students indicated prior participation and 11 indicated that they had not. In the post-participation survey, five students indicated that they had participated beyond the current semester and six indicated that this was their first semester of participation. Two of the post-participation respondents indicating prior participation indicated a duration of prior participation of five semesters; the remainder indicated a period of participation of two semesters. In the pre-participation survey, four students indicated one semester of participation, seven indicated two semesters, two indicated three semesters, and two indicated four semesters of participation. Of the post-participation respondents, six participated for academic credit during the semester (the pre-participation responses were gathered prior to the introduction of one form of participation for credit during the semester and thus cannot be compared to these post-participation numbers). The academic credit participants had a significantly higher retention rate (six of seven completing) as compared to the non-academic credit participants, as might be expected.

10.8.1 Data Collected

In both surveys, respondents were asked to indicate all of the benefits that they expected to receive (pre-participation survey) or had received (post-participation survey) in response to the eighth question. The list of possible choices presented is included below. Respondents were also given the opportunity to write in any additional areas of benefit that they expected/hoped to receive or had received. The initial list was generated through prior surveys, such as those presented in [3], and other anecdotal feedback.

Knowledge about spacecraft design	Experience working on a large group project
Knowledge about structured design processes	Experience with a structured design process
Knowledge about a particular technical topic	Experience related to a particular technical topic
Knowledge about project management	Project management experience
Knowledge about time management	Time management experience
Leadership experience	Improving leadership skills
Improving technical skills	Improving project management skills
Improving time management skills	Understanding of how my discipline relates to others
Experience working with those from other disciplines	Learn other discipline's technical details/terminology
Real-world project experience	Improved chance of being hired in desired field
Item for resume	Ability to present at professional conference
Improved presentation skills	Ability to present at professional conference
Inclusion as author on technical paper	Recognition in the university community

In Fig. 10.35, the benefits expected and received by respondents are presented. In 13 (of the 26 categories), a greater percentage of respondents expected the benefit than received it. In two cases, the level of those expecting and receiving the benefit was the same and in all other (11) cases, a higher percentage of individuals indicated receipt of the benefit than indicated expectation of receiving it.

The expectation and receipt of each benefit type by undergraduate versus graduate respondents is considered in Fig. 10.36. Note the high prevalence of high (100 %, of the two) benefit receipt, occurring in 16 of the categories, by the graduate students, in many cases exceeding the number of individuals expecting the given type of benefit. The percentage expecting a benefit exceeded the percentage receiving it in only eight cases, for undergraduates. One tie exists, and in all other cases, the percentage receiving the benefit exceeded those expecting it.

The benefits received by all students and students participating for course credit versus not participating for course credit are presented in Fig. 10.37. Note that only received benefits are presented, as there is no comparable data set for expectations of course versus non-course participants. In 16 of the 26 categories, those not participating for course credit have higher levels of the percentage receiving the bene-

10.8 Summative Evaluation Tools: Meeting Expectations Survey

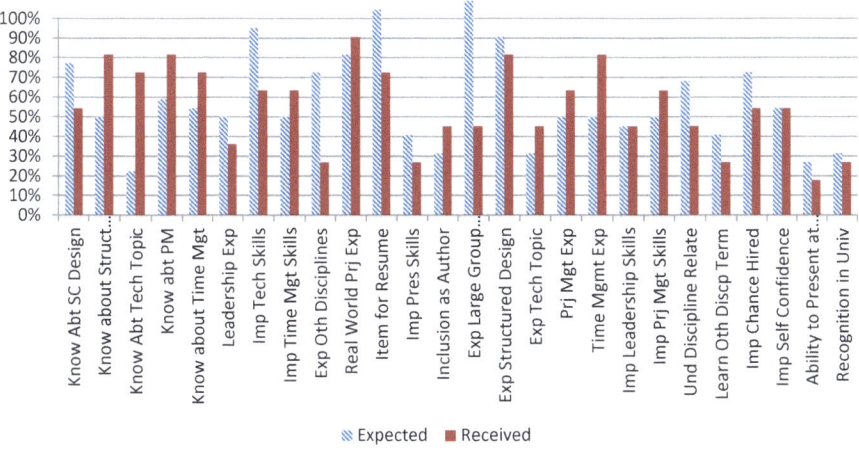

Fig. 10.35 Benefits expected and received by respondents

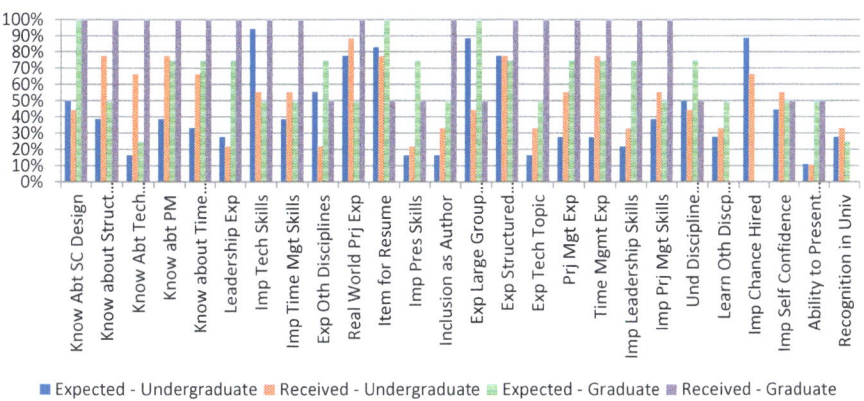

Fig. 10.36 Benefits expected and received, by graduate versus undergraduate status

fit; in those remaining, the participants for academic credit report higher percentages receiving the benefit. Of the students participating for academic credit, two-thirds were participating in the context of a project management class and thus would not have been exposed to several types of benefit causing activities that others (both for-credit and not-for-credit) participants were. One half of these students participated in the project above-and-beyond course requirements (receiving exposure to these benefits), while one-half did not.

The post-participation survey respondents were also (like those taking the pre-participation survey) asked to identify their reasons for participation. These are presented in Fig. 10.38. Note that no respondents indicated participation due to the participation of a particular faculty member while over half participated due to an

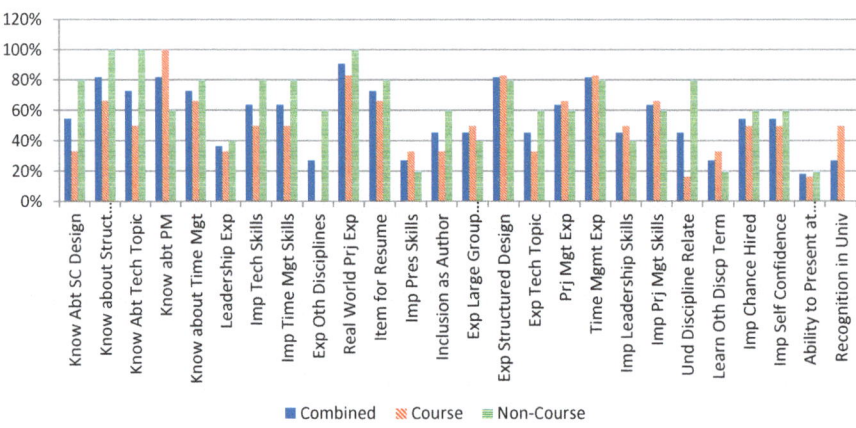

Fig. 10.37 Benefits received by students participating in a course versus participating for extra-curricular enrichment

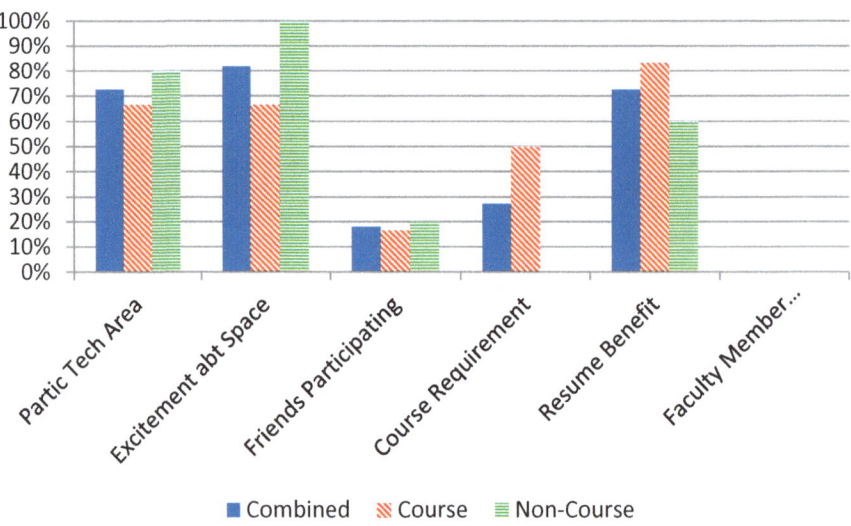

Fig. 10.38 Reasons identified for participating by post-participation survey respondents

interest in a technical area excitement about space and for a resume benefit (respondents could select as many responses in this area as they desired).

Respondents were asked to identify the top three benefits that they expected to receive and received from participation. These results are presented in Fig. 10.39 and Table 10.15. As shown in Fig. 10.39, three categories have individuals indicating them as a top expectation and no one indicating them as a top receipt and three have top benefits received identified with no corresponding identification as a top expectation. In six cases (excluding the foregoing three without any corresponding receipt), more individuals indicate an area as a top expectation than a top receipt.

10.8 Summative Evaluation Tools: Meeting Expectations Survey

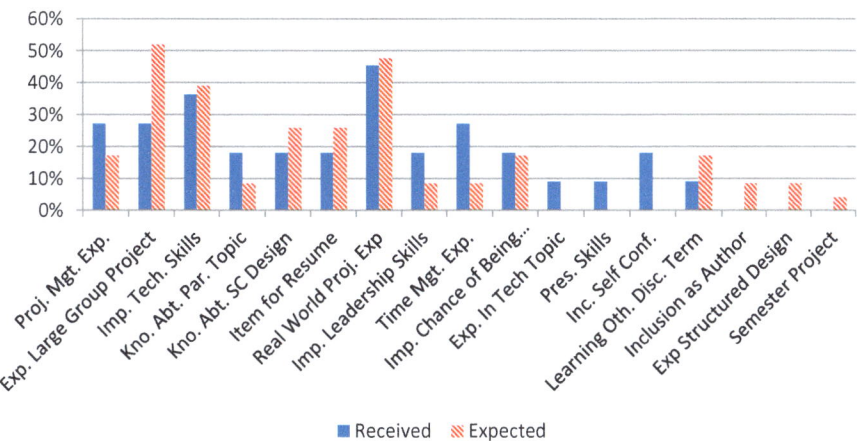

Fig. 10.39 Top benefits expected and received (note that similar categories have been combined)

The opposite is true for five categories. Table 10.15 provides a breakdown of the percentage of respondents choosing each benefit in the first, second, and third positions, as well as their cumulative percentage numbers.

Respondents were, finally, asked to identify whether they planned to seek employment in a field related to their area of participation and whether they believed that participation would help them secure employment. Respondents were asked to indicate a value on a nine-point scale ranging from 9-strongly agree (that they were seeking employment in a relate field or would be aided by participation in securing employment) to 7-agree, to 5-neutral, to 3-disagree to 1-strongly disagree. Respondents responses to this, including for the combined set of postsurvey respondents and separated into those participating for course credit (vs. not) and by graduate versus undergraduate status, are presented in Table 10.16. Note that all of the averages are neutral or above.

10.8.2 Analysis of Data

The data presented demonstrates significant attrition, particularly among those students who are not receiving academic credit for their participation. The attrition is likely attributable to several factors. Students may be interested in participating but lack time (on a progressively greater basis as the semester progresses and workload increases). They may leave because they are not receiving the benefits that they sought (which lead to their initial decision to participate). They may leave due to other program-attributable factors (such as not getting to work in a desired area, a lack of agreement with a direction taken, etc.) or to personal factors. The characterization of this attrition represents a subject for future work.

There are several areas where benefits underperform or overperform expectations. A difference of more than 20 %, with the percentage of those receiving the benefit lower than those expecting it, exists in the areas of knowledge about space-

Table 10.15 Top three benefits sought and received

	Received				Expected			
	1st (%)	2nd (%)	3rd (%)	Cumulative (%)	1st (%)	2nd (%)	3rd (%)	Cumulative (%)
Proj. mgt. exp.	9	0	18	27	13	4	0	17
Exp. large group project	18	0	9	27	17	22	13	52
Imp. tech. skills	9	9	18	36	35	4	0	39
Kno. abt. par. topic	18	0	0	18	0	4	4	9
Kno. abt. SC design	9	0	9	18	4	17	4	26
Item for resume	9	9	0	18	4	13	9	26
Real-world proj. exp	18	9	18	45	13	22	13	48
Imp. leadership skills	9	9	0	18	4	4	0	9
Time mgt. exp.	0	27	0	27	0	0	9	9
Imp. chance of being hired	0	9	9	18	0	0	17	17
Exp. in tech. topic	0	9	0	9	0	0	0	0
Pres. skills	0	9	0	9	0	0	0	0
Inc. self-onf.	0	9	9	18	0	0	0	0
Learning oth. disc. term	0	0	9	9	4	9	4	17
Inclusion as author	0	0	0	0	0	4	4	9
Exp. structured design	0	0	0	0	0	4	4	9
Semester project	0	0	0	0	4	0	0	4

Table 10.16 Are post-participation survey respondents seeking employment in their field of participation and do they believe that participation aided them in securing employment?

	Seeking employment in field	Aid in securing employment
Combined	6.3	7.1
Course	5.8	6.3
Non-course	6.8	8.0
Grad.	5.0	8.0
Undergrad.	6.6	6.9

craft design, improved technical skills, item for resume, experience working with other disciplines, and experience with a large group project. Conversely, more than 20% individuals received than expected to receive knowledge about the structured design project, knowledge about a technical topic, knowledge about project management, and time management experience. The number of areas of underperformance relative to expectations is similar for both graduate and undergraduate students, with graduate students having less percentage of individuals indicating receipt of a benefit expected than desiring it in seven cases versus eight for undergraduates. In all others (19 for graduates, 18 for undergraduates), the percentage of individuals receiving the benefit met or exceeded those desiring it.

The end of semester responses for seeking employment in a field related to participation and whether individuals believe that participation will aid in securing employment show a moderate decline. Initially, the average score was 7.14 (slightly above 7-agree); this has dropped to 6.3 (slightly below 7-agree). The response regarding the project aiding in securing employment had dropped from 7.33 to 7.1. While numerous causes may exist for these changes (e.g., students learning that they are not interested in the field, students gaining a better understanding of what more they need to learn to be successful in the field) neither represents enough movement relative to the scale to be practically significant. Both remain in the agreement range and in the same general area of this range. Notably, those not participating for course credit indicate a higher average score in both areas, potentially identifying this interest and belief in job placement benefit attainment as a reason for retention.

10.8.3 Summary

This section has presented work on the characterization of benefits that students believe that they are receiving from participating in a small spacecraft development program, relative to the benefits that they are seeking. The data presented shows that students who are retained in the program are receiving a similar level of benefits to those desired, but that the areas of benefit expected/desired and the areas of benefit received are not always the same. The characterization of the impact of benefit receipt on program attrition as well as the limited misalignment between benefits sought initially and benefits received is beyond the scope of this analysis and serves as an important topic for future work.

While the work presented supports the conclusion that students who participate for an entire semester are receiving benefits similar to those desired, no extrapolation is currently possible to the greater small spacecraft development community. Further analysis of the data from the nationwide survey (discussed in Chap. 9) may provide insight into this area.

10.9 Results of Prior Summative Evaluation

The summative evaluation work, described in the proceeding sections, has demonstrated the efficacy of small spacecraft development programs, and OOSDI in particular, for providing students with a variety of educational benefits. The work has also demonstrated the efficacy of several evaluation techniques for assessing student benefits from program participation. This assessment is, of course, difficult due to the unique experience that every student has from participation. There is no shared curriculum that participants can be tested on or assessed against. Instead, it is necessary to rely on participant characterization of their pre- and post-participation states (as the use of pre- and posttests would presume a constant scale, which may be altered by participants learning more about the subject and realizing the limited scope of their understanding due to this, among other factors) and their identification of the experiences, attitudes, and benefits that they have derived from participation.

10.10 Conclusion

This chapter has presented work on assessment for a small spacecraft program. It has presented information about the benefits attained (and other characteristics) of participants in the University of North Dakota's OOSDI program. It has also discussed how similar assessment—both formative and summative—can be used by other programs and the benefits that others can derive from its use.

References

1. Straub, J., and D. Whalen. 2013. Student expectations from participating in a small spacecraft development program. *Aerospace* 1(1): 18–30.
2. ———. 2014. Evaluation of the educational impact of participation time in a small spacecraft development program. *Education Sciences* 4(1): 141–154.
3. ———. 2013. An assessment of educational benefits from the OpenOrbiter space program. *Education Sciences* 3(3): 259–278.
4. Sadler, D.R. 1989. Formative assessment and the design of instructional systems. *Instructional Science* 18(2): 119–144.
5. ———. 1998. Formative assessment: Revisiting the territory. *Assessment in Education* 5(1): 77–84.
6. Black, P., and D. Wiliam. 2009. Developing the theory of formative assessment. *Educational Assessment, Evaluation and Accountability (Formerly: Journal of Personnel Evaluation in Education)* 21(1): 5–31.
7. Nicol, D.J., and D. Macfarlane-Dick. 2006. Formative assessment and self-regulated learning: A model and seven principles of good feedback practice. *Studies in Higher Education* 31(2): 199–218.
8. Straub, J. 2015. Initial results from the first national survey of student outcomes from small satellite program participation. Presented at AIAA Space and Astronautics Forum and Exposition 2015.
9. Mueller, R.P. 2011. Lunabotics mining competition: Inspiration through accomplishment. Presented at Earth and Space 2012@ Struction, and Operations in Challenging Environments.
10. 2009. *Ethnography & Evaluation Research*. http://www.salgsite.org.
11. May 2009. *Undergraduate Research Student Self-Assessment (URSSA) FAQs*. http://www.colorado.edu/eer/downloads/URSSA_FAQs.pdf.
12. Ayob, A., R.A. Majid, A. Hussain, and M.M. Mustaffa. 2012. Creativity enhancement through experiential learning. *Advances in Natural and Applied Science* 6(2): 94–99.

Index

A
Academic institution decision-making process, 62
Academic institution investigators, 58
Active Learning Project Sequence (ALPS), 130
Active learning techniques, 131
Aerospace development, TRL, 38
Aerospace engineering, PBL, 103
Agriculture, visible light remote sensing data, 28
AIAA Small Satellite Technical Committee, 8
Air Force's University NanoSat Program, 22
Altair 8800 satellite, 6, 40
Architecture Analysis and Design Language (AADL), 177
Area-specific technical skills, gaining, 184
Artifacts, 95, 96
Artificial satellites, 39
Assemblies, 93
Assessment, 138–143
    formative (*see* Formative assessment)
Assessment of small spacecraft development programs, 153–154, 156–165
    determining program value, 155–157
        experiment implementation, 157
        summative assessment process, 156
        URSSA mechanism, 156–157
    educational benefit assessment, 155
    and goals, 154–155
    importance of program, 167
    overview, 151
    PBL and EE, 152
    reporting value, 158–165
        comparison to other programs, 159–165
        local reporting, 158–159
        spacecraft development, 152–153
        tracking and reporting program value over time, 157–158
Attitude determination and control system (ADCS), 12

B
Base framework system, 45
Basic systems, technology, 28
Bespoke approach, 42, 65
Bloom's Taxonomy perspective, 132
Bona fide research, small spacecraft, 24
Build-from-scratch approach, 40, 44–46

C
Chalk and talk approach, 78
Classroom-based activities, risks for, 80
Collaborative mission for developing countries, 29
Communications approach, 91
Communications mission, 15
Communications satellite, 22
Conceive-Design-Implement-Operate (CDIO) process, 9, 25
Concept and architecture development process, 88–89
Conceptualization phase, program, 92–93
Construction and fabrication risks, 106
Costs, small spacecraft launch, 40
Creative Disciplines, 137
Creativity, 137
Critical path risk, 107
Crowdsourced innovation, 39

CSCI 297 class, students participating in, 156
CubeSat, 2, 6, 25, 38–42, 49, 132, 138
  bespoke approach, 42
  construction, 10
  cost, 3, 55
  development community, 23
  form factor, 11, 22, 23, 40, 55, 77, 153
  launch costs, 40
  low-cost, 39
  lower launch cost, 42
  OPEN framework, 42
  partially modified, 50
  programs, 4, 9, 155, 159
  projects, 103
  remote sensing mission, 14
  standard, small spacecraft, 21
CubeSat-based Space Missions for Geospace and Atmospheric Research program, 4
CubeSat-class spacecraft, 4, 9, 11, 43, 153
Current OPEN design, 30
Curriculum-integrated project, 130–131
Customer value chain analysis, 130

D
Damage and rework costs, 108–109
DARPA's SeeMe program, 3
Delft University of Technology, 130
Deliverables, 95, 96
Department of Defence (DoD), 42
Design frameworks, comparison of, 86–88
Design labor costs, 44
Design phase, program, 93
Design-from-scratch approach, 44
Develop-from-scratch approach, 47
Development phase, program, 93
Dickson satellite, 22
Disciple-agnostic instrument, 204
Dual-objective programs, 134–135
Dynamic Ionosphere CubeSat Experiment (DICE) program, 4

E
Economic model, 14
Ecosystem of innovation, 39
Educational benefits, 129
  aerospace engineering, 132
  ALPS, 130
  assessment, 155
  collaborative architecture, engineering and construction curriculum, 131
  curriculum-integrated project, 130
  educational program, 133–135
  information technology, 131
  inquiry-based approach to remote sensing, 130
  Internet, 131
  iterative engineering learning model, 132
  PBL (*see* Problem/project-based learning (PBL))
  project-based learning, 130
  quantitative survey-based data, 130
  STEM skills, 129
  team efficiency, 132
  unified engineering course, 131
  VSSEC, 129
Educational goals
  benefits, 129–133
  faculty research, 133
  nontraditional disciplines, 136–137
  risk, 133
  soft and Other skills, 137–138
  student involvement, 133
  technical discipline skills, 135–136
Educational Launch of NanoSats/Nanosatellites (ELaNa) program, 21, 40, 83, 159, 160
Educational programs
  benefit, 134
  definition, 133
  dual-objective programs, 134–135
  workplace-analog benefits, 134
Education (and/or various sciences), 137
Education-only program, 58–61
  benefits, 59–60
  drawbacks, 61
Educators, challenge facing, 78
Effective Media Relations: How to Get Results, 158
Eight-step approach, PBL, 25
Electrical power system (EPS), 15
ESA, 153
Estimation error, schedule, 107
European Student Earth Orbiter, 22
ExoplanetSat, initiative of, 25
Experiential education (EE), 54
  assessment of small spacecraft development programs, 152
  implementation of, 54
Experiential learning (EL), 24–26, 78, 80
Experimental space technologies, 38–39
Explorer One Prime Satellite, 41
Explorer 1 satellite, 39
Export Administration Regulations (EAR), 50, 58
Export control regimes, 30

Index 225

External partner research program, 56–58
　benefits, 57–58
　control perspective, 57
　drawbacks, 58
　environmental perspective, 57
　policies, 57

F
Factual learning, 80
Field-based/realistic-environment PBL, 103
Financial and other resources, spacecraft development, 67–68
Flagship course, 131
Fly Your Satellite, 83
Formal learning technique, 24
Formative assessment, 138–143, 175
　class status of undergraduate respondents, 145
　data collection, 145–146
　outcome, 146
　post-participation survey, 144
　pre-participation survey, 144
　small satellite program participation, 144
　students
　　benefits sought by participants, 140, 141
　　breakdown of student participants, 139
　　choices selected by respondents, 141, 142
　　design and development of small spacecraft, 138
　　expectations from program participation, 139
　　participant response, 141–143
　　participation, 139
　　pre-identified project goals, 140
　　responses, 140
　　survey, 138–140
　　university-class spacecraft, 138
　survey respondents, 145
Formative evaluation, 175–198
　gain quantification survey, 183–198
　　benefits of interdisciplinary projects, 183–184
　　data and analysis, 185
　　learning objectives, 184–185
　　results by amount of time participating, 193–194
　　results by GPA, 195–197
　　results by level of weekly participation, 191–192
　　results by undergraduate class level, 197–198
　　results of team leads *vs.* participants, 189–191
　　results of undergraduate *vs.* graduate students, 187–189
　interest/reason for participation survey
　　analysis of data, 182–183
　　data collected, 178–182
　　experimental design, 176–178
　　results and program enhancements in prior work, 198
　　tools, 176–183
Framework-based approach, 40, 42, 43, 65
　satellites, 39

G
Gain quantification survey, 176, 183–198, 200–203
　formative evaluation, 183–198
　　benefits of interdisciplinary projects, 183–184
　　data and analysis, 185
　　learning objectives, 184–185
　　results by amount of time participating, 193–194
　　results by GPA, 195–197
　　results by level of weekly participation, 191–192
　　results by undergraduate class level, 197–198
　　results of team leads *vs.* participants, 189–191
　　results of undergraduate *vs.* graduate students, 187–189
　summative evaluation, 199–204
Graduate respondents, 214
　average aggregate improvement, 188
　beginning status levels, 188
　survey, 176, 183
Ground station software team, 72
Ground systems, 90–91

H
Hardware costs, 47–48
　parts costs, 47
　testing materials and supplies Costs, 48
　waste and damage costs, 47
Hemispherical Anti-Twist Tracking System (HATTS), 24
Home grown technology, 31
Human resources, spacecraft development, 65–67
　student involvement, 66
　student risk perception, 66–67
Hybrid research education program, 61

I

Image processing techniques, 29
Industry-analog project, 131
Inexperienced workers, 101, 102
  cost of, 122
  types of, 118
  value model for, 122–123
  value proposition for use of, 122
Informal learning technique, 24
Inquiry-based approach to remote sensing, 130
Integration labor costs, 44–45
Integration testing, 93
Intelsat 10 satellite, 22
Interdisciplinary projects, benefits of, 26–27, 183–184
Internal research program, 55–56
  benefits, 56
  drawbacks, 56
  research question, 55
International Space Station, 67, 101
International Space Station Agricultural Camera, 12, 28
International Trafficking in Armaments Regulations (ITAR), 50, 58
Internet, 131
Interns, 124
Interorbital Systems, 3
Inter-spacecraft communication, 26
ION-1 (oxygen airglow photometer) spacecraft, 23
ION-2 (neutral hydrogen photometer) spacecraft, 23
Iterative engineering learning model, 132

J

Junior employees, 124

K

Kit-based approach, 40–41, 49
Knowledge Retention, 138

L

Lab equipment costs, 48
Labor costs, 44–47
  design Labor, 44
  integration labor, 44–45
  management labor, 46
  operations labor, 45
  testing and validation labor, 45
Larger-sized small spacecraft, 22
Launch and operations phase, program, 93–94
Learning technology, principles, 79, 80
Learning, value of, 80
Low-cost development approaches, 11
Low-cost test platform, 38
Lunabotics competition, NASA, 184

M

Magnitude-of-impact peril model, 82
Management labor costs, 46
Management/Project Management, 137
Marketing, 137
Massachusetts Institute of Technology, 130
Massachusetts Institute of Technology Aeronautics and Astronautics program, 131
Meeting expectations survey, 214–221
  analysis of data, 219–221
  data collected, 216–219
Mentor, 210
Miscommitment, students, 110
  change in course load, 117
  external work commitment/change, 116–117
  involvement in other university activities, 117
  underestimation of coursework, 116
Mission analysis and design, 88–90
  concept and architecture development, 88–89
  defining objectives, requirements, and constraints, 88
  drivers, requirements, analysis, and selection, 89–90
Mission architecture, defining, 90–91
  communications approach, 91
  elements, 89
  ground systems, 90–91
  payload elements and bus, 90
  subject, 90
  target altitude, orbit, and estimated mission duration, 90
Mission concept, 88
  creating and selecting, 85
Mission engineering phases, correlation of, 87
Mission objectives, 84, 86
Mission operations costs, 67
Mission-specific payload hardware components, 55
Mission-specific technologies, 28
Mitigation techniques, 110
Mix and match approach, 51
Morphological analysis, 130
Mosaicking technology, 13, 29

Index   227

Motivation, 137
Multi-frame super-resolution algorithms, 13

N
NanoSat/Nanosatellites, launching, 3, 37–39
  ecosystem of innovation, 39
  low-cost test platform, 38
  to mature technical readiness of experimental space technologies, 38–39
National Aeronautics and Space Administration (NASA), 42, 153
National space competency, considerations based on, 31
Near-space environment, 130
Near-space mission, 92
Near-visible light imagery, 27
Non-CubeSat-class spacecraft, 22
Non-PBL methods, 156
Nontraditional disciplines, 136–137

O
Objective definition process, 78–79
Object process methodology, 130
OOSDI program, 154, 155, 176, 184, 185, 198, 199, 215
Open Prototype for Educational NanoSats (OPEN), 153, 155
  design, 6
  framework, 42, 71, 73
OPEN-class 1-U, 155
OpenOrbiter, 155
  organizational structure, 70
  program, 60, 68, 73, 157, 212
  project, 156
  satellite, 71
  small spacecraft development program, 138
  spacecraft, 71
Operations labor costs, 45
Optical system group, 71
Orbital services model, 13
Orbiting Picosatellite Automatic Launcher (OPAL), 23
Organizational strategies for program formation, 68–73
  implementation difficulties, 72–73
    divergent interests of faculty participants, 73
    faculty support, 72
    logistical challenges, 73
    technical challenges, 73
  program implementation, 69–72
    communications, outreach and policy, 71
    electrical, 71
    ground station, 72
    mechanical, 71
    mission design and architecture, 69
    operations, 72
    software, 71

P
Paid student workers, 124
Parts costs, 47
PBL/EE projects, 78
PBL/EE-driven course, 54
Personal satellite, 6–7
Planning process, 94, 95
PocketQub form factor, 23
Policy, 137
Poly-PicoSatellite Orbital Deployed (P-POD), 23
Post-participation survey, 214
  respondents, 217
  respondents seeking employment, 181, 220
  student respondents for, 215
Pre-participation survey, 214
  student respondents for, 215
Problem/project-based learning (PBL), 9, 24–26, 54, 78, 102–103, 130, 156
  activities, 133
  assessment of small spacecraft development programs, 152
  benefits, 25, 132
  classification system, 130
  eight-step approach, 25
  environment, 137
  examples, 132
  field-based/realistic-environment, 103, 132
  formats, 25
  implementation of, 54
  incorporation, 130
  instruction of particular material, 132
  knowledge retention, 138
  and robotic Mars mission, 129
  scientific-technological focus, 54
  self-image, 138
  soft skills, 132
  student retention, 138
  subject disciplines, 132
  subject-specific skills, 132
  synchronous and asynchronous elements, 131
  techniques, 132

Problem/project-based learning (PBL) (*cont.*)
  type, 132
  use of, 24
  utility, 132
Program-attributable factors, 219
Prospective learning benefits, 80
Pugh Concept Selection, 130
Pumpkin Incorporated, 5

Q
Qualitative analysis, 29–31
  application suitability, 30
  mission approach considerations, 30–31
  quality and utility of data, 29–30
Qualitative evaluation of value of approaches, 42–50
  hardware costs, 47–48
  lab equipment costs, 48
  labor costs
Quality control processes, 106
Quality function deployment, 130
Quality management, 96
Quantitative survey-based data, 130

R
Remote sensing, 11
Remote sensing benefits, small spacecraft, 27–28
  agriculture, 28
  urban areas, 28
Requirements mapping technique, 96
Research Development Test and Evaluation (RDT&E), 38
Research-focused projects, 61
Risk model, 83
Robust approach, 153
Robust kit, 40
Robust risk management mechanisms, 67
Root cause analysis (RCA) techniques, 110–117
  inexperience symptoms, 112–114
    lack of attention to detail, 112–113
    lack of self-motivation, 113
    overconfidence causes failure, 113–114
    problem with work environment, 114
    unsure to perform task, 113
  miscommitment
    change in course load, 117
    external work commitment/change, 116–117
    involvement in other university activities, 117
    underestimation of coursework, 116
  scheduled turnover, 115–116
    end of course project period, 116
    end of paid work period, 116
    graduation, 115–116
  unscheduled turnover, 114–115
    departure from university, 115
    medical/family/personal problem, 115
    student takes internship, 115
    student transfers program/school, 114
    turnover due to miscommitment, 114

S
Satellite Quick Research Testbed project, 22–23
Satellites *vs.* early computers, 5–6
Scale-Invariant Feature Transform (SIFT), 13
Scenario graphing, 130
Schedule risk, students involvement
  critical path risk, 107
  schedule creep, 107–108
  schedule estimation error, 107
Scope creep, 108
Self-image, 138
Simplified mission analysis and design (SimplMAD) process, 86, 88
Single-category approach, 62
Single craft approach, 30
Single craft missions, 30
Skill learning, 80
SMAD (SMAD 4), 86
Small satellite, 144
Small satellite development programs, 81
Small spacecraft, 2–11, 13–14, 39–40, 101, 138–143
  benefits of, 9–11
    changing small satellite environment, 10
    space research, 10–11
    STEM education and small satellites, 9–10
  communications mission, 15
  education (*see* Educational goals)
  formative assessment (*see* Formative assessment)
  mission cost model, 68
  overview, 1–8
    academic proliferation, 4
    access to, 2
    comparison of capabilities, 7
    comparison of satellites and early computers, 5–6
    mainstream, 3–4
    nanosatellites, 3
    personal satellite, 6–7

Index   229

    start of industry proliferation, 5
    the status quo, 2–3
    towards future, 7–8
  technologies and missions, 12–14
    example remote sensing
      mission, 14
    super-resolution and mosaicking, 13
    task sharing between craft, 13–14
  types of, 8–9
  uses of, 11–15
Small spacecraft development programs,
    24–29, 31, 38–39, 43–50, 55, 59,
    66–73, 78–88, 91–94, 151
  approaches, 39–40
  assessment of (*see* Assessment of small
    spacecraft development programs)
  assurance, 96
  bespoke approach, 42
  costs of, 55
  creating and selecting mission concept, 85
  design framework and level of program
    rigidity, 86–92
    comparison of design frameworks,
      86–88
    defining mission architecture (*see*
      Mission architecture, defining)
    driver identification, 91
    mission analysis and design (*see*
      Mission analysis and design)
    requirements, analysis, and selection,
      91–92
  educational benefits, 24–27
    experiential and problem-based
      learning, 24–26
    of interdisciplinary projects, 26–27
  financial and other resources, 67–68
  framework-based approach, 42
  goal-based technique, 84–86
    constraints, 85
    defining requirements and constraints,
      84–85
    functional and operational
      requirements, 84
  human resources, 65–67
    student involvement, 66
    student risk perception, 66–67
  identifying science, technology
    development, educational, and other
    goals, 78–83
    defining objectives, 78–79
    maximizing value, 79–80
    value assessment, 80–83
  kit-based approach, 40–41
  launching NanoSat, 37–39

    ecosystem of innovation, 39
    low-cost test platform, 38
    to mature technical readiness of
      experimental space technologies,
      38–39
  matching goals and funding sources,
    83–84, 160
  mix and match approach, 51
  organizational strategies for program
    formation (*see* Organizational
    strategies for program formation)
  overview, 21–22
  planning for program longevity, 92–94
    closeout, 94
    conceptualization, 92–93
    design, 93
    development, 93
    launch and operations, 93–94
  processes for mission management, 94–96
  project/mission management, 94–95
  qualitative evaluation of value, 42–50
    allowing focus on area of interest, 50
    benefits related to export control (EAR/
      ITAR), 50
    consideration of recurring amortized
      vendor development costs, 48–49
    cost levels, 43–48
    ease of modification and extensions of
      design, 49–50
  reasons for forming program, 77–78
  research benefits, 22–24
  societal benefits, 27–31
    collaborative mission for developing
      countries, 29
    considerations based on national space
      competency, 31
    qualitative analysis (*see* Qualitative
      analysis)
    remote sensing benefits, data products,
      and their uses, 27–28
    technologies and mission, 28–29
  systems and processes, 95–96
  using objectives, requirements, and
    constraints for decision making, 86
Social Media: Dominating Strategies for
  Social Media Marketing with
  Twitter, Facebook, Youtube,
  LinkedIn, and Instagram, 158
Soft and other skills, 137–138
Software developers, 39
Software technologies, 13, 28
Spacecraft
  costs, 68
  reducing size of, 55

Spacecraft Systems Engineering, 4th Edition (SSE 4), 86
Space Mission Architecture and Design, 3rd Edition (SMAD 3), 86
Space research, 10–11
Speeded Up Robust Features (SURF), 13
Sputnik 1 satellite, 39
Sputnik, small satellite, 22
SQUIRM/SQUIRM-E model, 119, 123
SQUIRM-Extended Model (SQUIRM-E), 110, 111, 117–120, 123, 124
Stakeholders, 79, 158
STEM skills, 129, 132, 136
Student-involved projects, 26
Student involvement and risk, 104–110, 112–123
　application, 119–120
　to faculty research, 103
　future work, 124–125
　interns, 124
　junior employees, 124
　overview, 101–102
　paid student workers, 124
　PBL, 102–103
　quantifying model, 120–121
　　combining for result, 121
　　data for model parameters, 121
　　mitigation/response assessment, 121
　　risk assessment, 120–121
　RCA techniques, 110–117
　　inexperience symptoms, 112–114
　　miscommitment, 116–117
　　scheduled turnover, 115–116
　　unscheduled turnover, 114–115
　risk model (*see* Student qualitative undertaking involvement risk model)
　risk perception, 103–104
　student volunteers, 123
　using SQUIRM *vs.* SQUIRM-E, 117–119
　　choosing model, 119
　　comparative simplicity, 118
　　particulars of student work environment, 118–119
　　project size, 118
　　types of inexperienced workers, 118
　value model for inexperienced workers, 122–123
　　cost, 122
　　discontinuous innovation benefits, 123
　　training benefits, 122–123
Student qualitative undertaking involvement risk model, 104–110
　cost risk, 108–109
　　buying time, 109
　　cost creep, 108
　　cost estimation error, 108
　　damage and rework, 108–109
　　risks posed by student worker involvement, 109–110
　　inexperience, 110
　　miscommitment, 110
　　scheduled turnover, 109
　　unscheduled turnover, 110
　schedule risk, 106–108
　　critical path risk, 107
　　schedule creep, 107–108
　　schedule estimation error, 107
　technical risk, 105–106
　　component, 106
　　construction/fabrication, 106
　　integration, 106
　technical, schedule and other standard risks, 104–105
Student Qualitative Undertaking Involvement Risk Model (SQUIRM), 67, 110
　framework, 102–105, 117–119, 123, 124, 133
　vs. SQUIRM-E, 117
　　choosing model, 119
　　comparative simplicity, 118
　　particulars of student work environment, 118–119
　　project size, 118
　　types of inexperienced workers, 118
Student retention, 138
Students in interdisciplinary projects, 27
Student Space Exploration and Technology Initiative, 22
Student volunteers, 123
Summative evaluation, 199–203, 205–214, 216–221
　gain quantification survey, 199–204
　meeting expectations survey, 214–221
　　analysis of data, 219–221
　　data collected, 216–219
　results of, 221–222
　URSSA, 204–214
　　experiment implementation, 205–214
　　mechanism, 205
Super-resolution technology, 13, 29
Survey
　academic level of respondents, 161
　academic year of respondents, 212
　belief in participation aiding employment, 165
　benefits expected and received by respondents, 217, 219
　benefits sought by participants, 179
　choices selected by respondents, 180

Index

confidence, comfort, and skills gained, 207
hours spent, 213
improvement of technical skills, 165
information about respondents, 159–161
interest in employment in field of participation, 164
outcomes, 167
participant gains from participating, 164
participation increased interest in space, 166
participation increased leadership skills, 166
participation increased project management skills, 166
post-participation, 214
post-participation survey respondents, 218
preparation benefits of participation, 211
pre-participation, 214
reasons for participation, 163, 181, 213
respondent attitudes, 210
respondent class for undergraduates, 161
respondent GPA, 162
respondent weekly commitment, 163
respondent years in current academic program, 161
respondents participation, 161–165
semesters involved in program, 177, 201, 202
skills and abilities gained, 208
skills gained, 206
sources of information about program, 213
student accomplishments, 211
undergraduate and graduate respondents, 176, 183, 214
System level testing, 93

T
Task sharing between craft, 13–14, 29
Team leads and participants
 average aggregate improvement, 190
 beginning status level, 190
 improvement in status levels, 191
Technical discipline skills, 135–136
Technical risk, students involvement
 component, 106
 construction/fabrication, 106
 integration, 106
Testability of requirement, 84
Testing and validation labor costs, 45
Testing materials and supplies costs, 48
Time management learning, 154
Trade analysis process, 89, 91
Traditional style approach, 82

TRL 10 paradigm, 23
Tyvak Nano-Satellite Systems Company, 3, 5, 40

U
UAV, 79
1-U CubeSat, 7, 41, 83
Undergraduate Research Student Self-Assessment (URSSA), 159, 204–214
 instrument, 199
 mechanism, 156–157
 summative evaluation, 204–214
  experiment implementation, 205–214
  mechanism, 205
Undergraduate respondents, 214
 average aggregate improvement, 188
 beginning status levels, 188
 class status of, 215
 survey, 160, 176, 182, 183
Unified engineering course, 131
United Launch Alliance (ULA), 59, 83
Unit testing, 93
University-class missions, 8
University-class satellites, 23
University-class spacecraft, 21, 22
University-developed spacecraft, 30
University NanoSat program, 83, 159, 160
University of North Dakota's OpenOrbiter program, 144
Unplanned benefit, 134
Urban areas, remote sensing, 28
U.S. Air Force, 153
The U.S. National Science Foundation, 4
Utility analysis, 89, 92

V
Value assessment process, 80–83
Van Allen Radiation Belt, 41
Vendor amortized cost, 48
Vendor kit approach, 44–47, 65
Victorian Space Science Education Centre (VSSEC), 129
Visible light imagery, 27
Visible light remote sensing data, 11, 27, 28
 for agriculture, 12, 28
 for urban areas, 12, 28
V-model approach, 130

W
Waste and damage costs, 47